Pesticides

Pesticides

Health, Safety and the Environment

Second Edition

G.A. Matthews

Emeritus Professor, Department of Life Sciences,
Imperial College, UK

WILEY Blackwell

This edition first published 2016 © 2016 by John Wiley & Sons, Ltd
First edition published 2006 by Blackwell Publishing Ltd

Registered Office
John Wiley & Sons, Ltd, The Atrium, Southern Gate, Chichester, West Sussex, PO19 8SQ, UK

Editorial Offices
9600 Garsington Road, Oxford, OX4 2DQ, UK
The Atrium, Southern Gate, Chichester, West Sussex, PO19 8SQ, UK
111 River Street, Hoboken, NJ 07030-5774, USA

For details of our global editorial offices, for customer services and for information about how to apply for permission to reuse the copyright material in this book please see our website at www.wiley.com/wiley-blackwell.

Library of Congress Cataloging-in-Publication Data

Matthews, G. A., author.
Pesticides : health, safety and the environment / G.A. Matthews. – Second edition.
pages cm
Includes bibliographical references and index.
ISBN 978-1-118-97586-2 (cloth)
1. Pesticides. 2. Pesticides–Environmental aspects. 3. Pesticides–Health aspects. I. Title.
SB951.M5127 2015
628.5′29–dc23

2015018493

A catalogue record for this book is available from the British Library.

Wiley also publishes its books in a variety of electronic formats. Some content that appears in print may not be available in electronic books.

Cover image: Cover Image © Eric Hislop

Set in 10/13pt Palatino by SPi Global, Pondicherry, India
Printed and bound in Singapore by Markono Print Media Pte Ltd

1 2016

Contents

Preface to the first edition

Pesticides have undoubtedly helped to increase agricultural production and control vectors of disease over the last five decades, but there has been increasing criticism since Rachel Carson alerted users to the side effects of some pesticides in the environment. My own involvement dates back to before Rachel Carson's book *Silent Spring* as I was with a team of entomologists seeking to control insect pests of cotton in Africa. We recognised then that pesticides should only be used in conjunction with other control tactics, a system recognised in the USA and much publicised as integrated pest management. In the UK authorities had also responded to early problems due to use of highly toxic pesticides and the adverse effects of birds due to organochlorine insecticides by establishing the voluntary Pesticides Safety Precaution Scheme. While developed countries introduced registration of pesticides, requiring detailed scientific data on which to base a risk analysis, many other countries did not have the resources needed to operate a detailed registration system. In consequence highly toxic pesticides have been used in many countries, especially in tropical areas where protective clothing, as used in temperate climates, is unacceptably hot and uncomfortable to wear. This has led to many cases of illness and death following exposure to these highly toxic pesticides. These problems have been increasingly recognised and efforts made to harmonise registration requirements. This book sets out to emphasise that apart from correct choice of pesticide, it is the way it is applied that impacts on people, either directly on those using the many commercial products, but also others by the movement of pesticides in the environment and as residues in harvested produce.

Application technology has largely been ignored, and it has been left to engineers to design machinery that is easy to use and is as inexpensive as possible for the user. It is, however, a complex multidisciplinary subject which affects us all. Following the previous book, *Pesticide Application Methods*, that dealt with the different equipment that can be used, this book explains how the registration process can avoid use of the pesticides that pose a significant risk to users and the environment, and how by a better understanding of the subsequent movement of pesticides following application, the risk of any adverse impact following their use can be minimised. Today carefully applied pesticides, used only when needed, can contribute to higher productivity and allow us to feed and protect the growing human population. This requires much better education and practical training with certification so that pesticides are indeed applied more accurately and with greater safety than in the past. It is hoped that this compilation of data will help readers to have a better understanding of how pesticides can be applied without harming the users and adverse pollution of their environment. In addition, the overall management of pesticides, covering packaging, storage and proper stock control, needs to be improved so to avoid having obsolete stocks of pesticides. Unfortunately many countries still have obsolete pesticides that need careful disposal to avoid pollution of the environment.

There is a vast amount of information that has been published in scientific journals and books, so only selected data have been used in writing the chapters. More information is now available by the Internet, both from official web sites of government agencies and agrochemical companies, but also from pressure groups. However care is needed in choosing appropriate sources of information as sometimes only part of a story is reported. As with many complex subjects these days, it is important that as holistic approach as possible is made to obtain the benefits of the technology while minimising adverse effects.

Preface to second edition

Since publication of the first edition, much has happened in relation to registration of pesticides, especially in Europe. As legislation now takes greater note of the environment, emphasis has been switched from a scientific risk assessment of pesticides to use of the 'Precautionary Principle,' based on a hazard assessment. While some aspects of this might be welcome as it has removed many of the older, highly toxic insecticides, it has led to many unforeseen consequences. Farmers are particularly concerned about the loss of very important pesticides. One example is the ban on the use of the neonicotinoids as a seed treatment, which provided efficient control at a low dose and reduced the number of subsequent spray applications on the crop. This decision taken as a precaution was not supported by the scientific evidence, most clearly shown by the continued use of these insecticides in Australia where bees were not suffering from Varroa mite and other diseases spread by the mite.

Loss of some pesticides also reduces the range of chemicals with different modes of action needed in resistance management strategies. With environmental issues more prominent, the cost of bringing new pesticides to market will inevitably increase as more detailed evaluation of their impact on non-target species is required. Nevertheless new pesticides will undoubtedly be found by innovative chemistry, but registration will be a much more difficult process. Concurrently advancements in genetic engineering should enable plant breeders to select cultivars more resistant to pests and diseases so over time pesticides will be used as a 'last resort' in integrated pest management programmes and fulfil the new EU policy.

As the human population continues to increase, protection of crops from weeds, pests and diseases will continue to be crucial to have sustainable agriculture to feed the world. In this endeavour, pesticides will play a role, but in many countries greater efforts will be needed to improve training on all aspects in the life cycle of a pesticide product. Hopefully this revised edition effectively updates information to help achieve a greater understanding of their role.

Since the manuscript went to press, the UK Government changed the name of the Advisory Committee on Pesticides and it is now the Expert Committee on Pesticides.

Acknowledgements

I am grateful for new data on pesticides, which were provided through Crop Life International by Phillips McDougall. I wish to thank Keith Walters who kindly reviewed Chapter 6 and the following for providing the new illustrations: Roy Bateman, Martin Baxter (TeeJet), John Clayton (Micron), Jerry Cross (East Malling Research), Richard Garnett (Wisdom Systems) Ken Giles (UCDavis), Richard Glass, Tobias Erick Hüni (Birchmeier) and Andreas Mesch (Gloria).

I thank Eric Hislop, formerly at the Long Ashton Research Station, for the photograph of orchard spraying on the front cover showing a standard axial fan sprayer that has been extensively used in orchards.

Many of the illustrations remain the same as in the first edition in which I thanked the following people who provided photographs/diagrams – Bill Basford, Roy Bateman, FERA, formerly the Central Science Laboratory, John Chandler (Exocet), Alison Craig Vincent Fallon, (whose photograph was supplied by Georgina Down), Greg Doruchowski, Benedict Gove, Franklin Hall, Hans Ganzelmeier, Hardi International, Professor Brian Hoskins (Reading University), Househam Sprayers, Kilgerm Group Ltd, Kathy Lewis, Micron Sprayers Limited, Paul Miller (Silsoe Research Institute), Philippa Powell (Royal Commission on Environmental Pollution), Ryszard Saluda and Jan Van de Zande.

A special thank you to Moira for her understanding while writing this edition.

1 Pesticides and agricultural development

Agriculture is now confronted with considerable pressure to increase production to feed a higher human population, expected to rise to 9 billion by 2050. In the Green Revolution of the late twentieth century, yields were increased by growing new crop varieties with more fertilizer and protected using pesticides. Growing concerns about the extensive use of pesticides has led to a policy, especially within Europe, of using pesticides only as a last resort in integrated pest management (IPM) programmes. IPM has been adopted initially in protected environments in which biocontrol and other non-chemical techniques have been effective on high-value fresh fruit and vegetable crops. This has to some extent been driven by consumers and supermarkets wanting produce without pesticide residues. In arable crops, the initial use of IPM has been with using economic thresholds to determine when to spray and promoting some biological control by management of field margins.

At present, farmers continue to regard pesticides as an essential tool to ensure that they can maintain production of crops of quality and quantity. Prior to the development of the modern pesticide industry, farmers had to rely very much on crop rotations and mechanical weed control with hoes, hoping for a good dry spell of weather so that the weeds dried and were not merely moved. They also hoped that insect pests and disease control could be ameliorated by choosing a good crop variety, which had some resistance to pest damage. When reconciling pesticide reduction with economic and environmental sustainability in arable farming in France, Lechenet et al. (2014) failed to detect any positive correlation between pesticide use intensity and both productivity (when organic farms were excluded) and profitability. This is not surprising as pesticides protect a crop from yield loss and do not increase the 'potential' yields, determined by soil fertility, rainfall and choice of crop variety.

Global estimates of crop losses due to insect pests, diseases caused by various pathogens and competition from weeds, vary depending on the crops involved and local variations in pest severity, but losses from 26 to 40% for major crops, with weeds causing the highest potential loss are

Pesticides: Health, Safety and the Environment, Second Edition. G.A. Matthews.

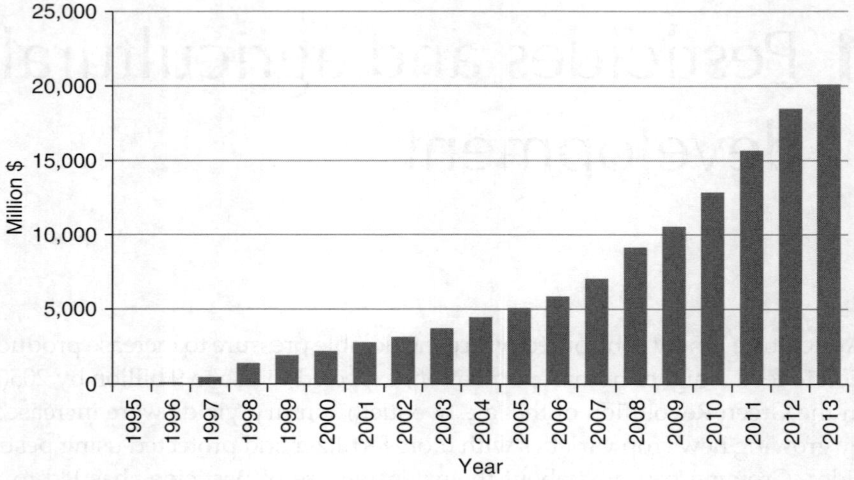

Fig. 1.1 Global market increase for GM seeds (Phillips McDougall, 2014).

reported (Oerke and Dehne, 2004). Some consumers have expressed a desire for 'organic' produce, but these usually are marketed at a higher price as farmers get lower yields and poorer quality without adequate crop protection. IPM has made considerable progress, especially within protected crops in which emphasis is given to cultural and biological control of pests, with minimal use of pesticides. Meanwhile the development of genetically modified crops has resulted in 18 million farmers growing biotech crops in 27 countries in 2013 covering 175.2 million hectares (Fig. 1.1). While insecticide use has reduced significantly on crops incorporating the Bt toxin effective against young larval instars of certain important pests, herbicide sprays continue to be needed with herbicide-tolerant crops.

Botanical insecticides, such as the pyrethrins, nicotine and rotenone (derris) were available prior to 1940, but they were not widely used, largely because they deteriorated rapidly in sunlight. A few inorganic chemicals, notably copper sulphate, lime sulphur and lead arsenate were also available. However, it was the development of synthetic organic pesticides during and following World War II that revolutionised the control of pests. Chemists had been looking for a cheap chemical with persistence in sunlight and low toxicity to man that would kill insect pests quickly, and in 1938 Muller showed that DDT would indeed fit this specification. Its availability during World War II led to initial use as a 10% dust on humans, for example in Naples, to suppress a typhus outbreak (Crauford-Benson, 1946). Soon afterwards it became available for agricultural use and began to be applied extensively on crops, such as cotton, at rates up to 4 kg ai/ha. Its use has had a major impact on vector control, being responsible, for example in India, for reducing the annual death rate due to malaria from 750,000 to 1,500 in

Table 1.1 Year of introduction of selected pesticides (Ware, 1986; MacBean, 2012 and web pages)

Year	Pesticide type	Pesticide
1850	Herbicide	Ferrous sulphate
1882	Fungicide	Bordeaux mixture
1930	Herbicide	DNOC
1931	Fungicide	Thiram
1939	Insecticide	DDT (commercialised 1944)
1942	Herbicide	2,4-D
1943	Fungicide	Zineb
1944	Insecticide	HCH (lindane)
1946	Insecticide	Parathion
1948	Insecticide	Aldrin, dieldrin
1949	Fungicide	Captan
1952	Insecticide	Diazinon
1953	Herbicide	Mecoprop
1955	Herbicide	Paraquat (commercialised 1962)
1956	Insecticide	Carbaryl
1965	Nematicide	Aldicarb
1968	Fungicide	Benomyl
1971	Herbicide	Glyphosate
1972	Insecticide	Diflubenzuron
1973	Insecticide	Permethrin
1990	Insecticide	Imidacloprid
	Fungicide	Azoxystrobin
	Insecticide	Spinosad
1994	Insecticide	Dicyclanil
	Fungicide	Ipconazole
1996	Insecticide	Pyriproxifen
1999	Fungicide	Ethaboxam
	Herbicide	Flucarbazone sodium
2001	Insecticide	Chlorfenapyr
2002	Insecticide	Pyridalyl
	Herbicide	Mesotrione
2003	Acaricide	Acequinocyl
2005	Insecticide	Flonicamid
	Insecticide	Spinetoram
	Insecticide	Spirodiclofen
2007	Insecticide	Chlorantraniliprole
	Insecticide	Flubendiamide
	Insecticide	Spirotetramat
	Herbicide	Pyraxosulfone
2012	Insecticide	Sulfoxaflor
	Insecticide	Flupyradifurone

the first eight years it was applied. Recognition of problems associated with the persistence of DDT in the environment were only realised later and highlighted by Rachel Carson in her book *Silent Spring* (Carson, 1962).

Parallel with the new insecticides, the development of 2,4-D as a herbicide controlling broadleaved weeds in cereal crops made a similar major impact on agriculture. While copper fungicides had been available since the end of the nineteenth century (Lodeman, 1896), further research has led to a greater range of more selective fungicides. These discoveries (Table 1.1) led to a rapid development of many other pesticides over the following decades. *The Pesticide Manual* (MacBean, 2012) is one important source of information on currently manufactured pesticides. Individual countries have lists of products that are registered for use. In the UK, this is published annually as *The UK Pesticide Guide*. Information can be obtained also from a number of Internet sites using a search engine such as Google. Information relevant to the UK is available through the Chemicals Regulation Directorate (CRD) web page, while the Environmental Protection Agency provides similar information in the USA. The Pesticide Action Network (PAN) and many universities also have web pages with pesticide information.

In Western Europe and North America the availability of herbicides was a major breakthrough at a time when shortages of labour due to the world war, industrialisation and urbanisation all played a part in necessitating a change in weed management on farms. Spraying fields with a herbicide allowed the crop seeds to germinate and develop without competition from weeds, thus increasing the harvested yield that could then also benefit from fertilizer applications. The discovery of paraquat, a herbicide that killed all weeds enabled a new concept of minimum or zero-tillage to reduce the need for ploughing fields every year and thus reduce the risk of soil erosion in many areas of the world. Subsequently glyphosate has dominated the weed control market with the advent of herbicide-tolerant genetically modified crops. Tolerance to other herbicides is now becoming available as overuse of one herbicide has led to weeds resistant to glyphosate.

The global market for pesticides has continued to grow despite the withdrawal of many of the older and more toxic pesticides, especially in Europe. Total global sales of pesticides had increased to approximately $54 million per annum in 2013, from $26.7 million in 2003. Growth of 9.8% per annum was recorded between 2007 and 2013 (Fig. 1.2a and b).

Data for 2013 shows that 43.7% of the global market was for the herbicide, plant growth regulator and sprout suppressants, with 27.5% for insecticides, 25.8% fungicides and 3.0% for other chemicals. North America was no longer the largest market, with Latin America, Asia and Europe all taking about 25% of the market (Fig. 1.3). Countries in Africa and the Middle East have not invested in increasing crop protection so yields remain low and crop protection has involved few pesticides and sowing genetically modified crops has barely started in these areas.

Few countries have survey data on the actual usage of pesticides. Thomas (2000) described the system operated in the UK to obtain accurate and timely information to satisfy government legislation. The data are also helpful in relation to the registration process and review of approved products.

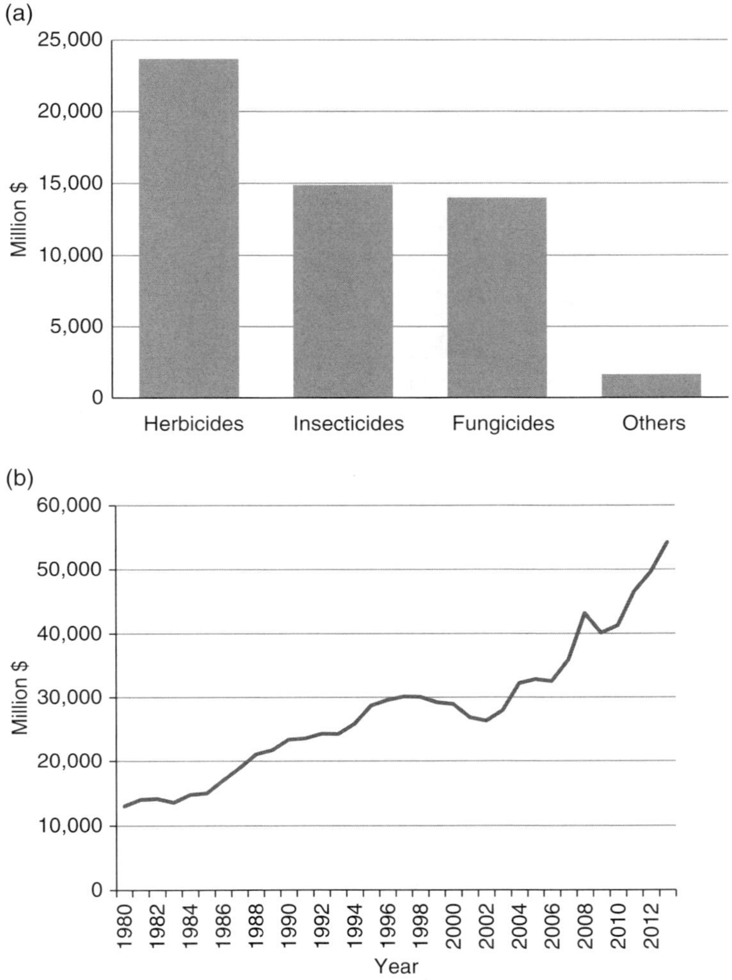

Fig. 1.2 (a) Global market of pesticides 2013 (Phillips McDougall, 2015). (b) Increase in global market of pesticides 1980–2013 (Phillips McDougall, 2015).

Principal pesticides

The following sections provide a brief account of some of the pesticides now available. World Health Organization (WHO) provides a classification of pesticides by hazard (WHO, 2010). Many of the pesticides which are extremely hazardous to use are no longer registered. This is particularly important in countries where protective clothing is uncomfortable to wear due to a hot climate. Poisoning due to pesticides is included in a classification of acute poisoning (Thundiyil et al., 2008). A major concern is that many more pesticides will be withdrawn in Europe when they are

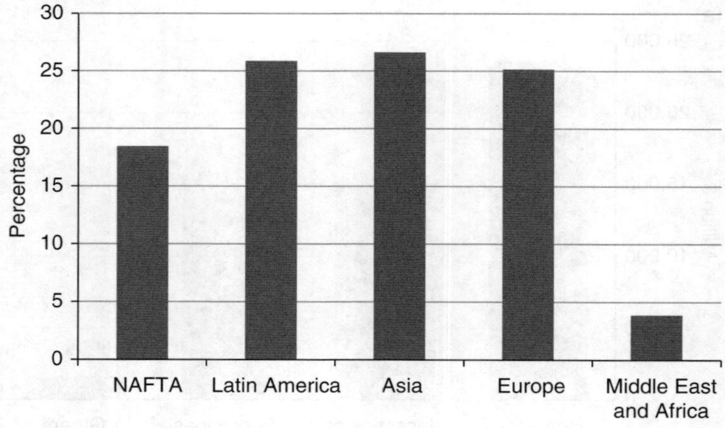

Fig. 1.3 Global market of pesticides by region 2013 (Phillips McDougall, 2015).

re-assessed with the hazard assessment due to other legislation such as the water directive. One report has predicted that a loss of certain pesticides, mainly on the basis of being perceived to be endocrine disruptors or the impact of the Water Framework Directive and greater reliance on mechanical and hand weeding, will among other consequences increase the cost of food at a time when agricultural output should be optimised (The Andersons Centre, 2014).

Insecticides

Initially the two main types of insecticides were the organochlorine (OC) and organophosphates (OPs), both being neurotoxins. The OC insecticides, including DDT, dieldrin and endrin had one main advantage, namely their persistence that enabled farmers to achieve control over a long period. However, plant growth and rainfall reduced the effectiveness of deposits on foliage. Later it was realised that this attribute led to residues remaining in the environment and being accumulated in some animals at the end of food chains. In consequence, these chemicals can be found everywhere, although their use has now been banned. The use of DDT for indoor residual spraying in vector control has been permitted on a limited scale.

OP insecticides are a diverse group (Anon and Committee on Toxicity of Chemicals in Food, Consumer Products and the Environment, 1999), some of which are extremely toxic, for example parathion, methidathion and monocrotophos, while others, such as temephos, malathion and trichlorfon are much less hazardous to use. When used in place of the OC insecticides, more people suffered acute poisoning, as the need for protective clothing had not been adequately recognised in many countries. Many people now consider that those classified as the most hazardous to use (see later), should also be banned. In the UK most of these chemicals were not approved, although a few, such as chlorfenvinphos, were registered for control of specific pests. Other OPs such as diazinon were used extensively in sheep dips. Karalliedde et al. (2001) provide a critical review of OPs and their impact on health.

Another group with a similar mode of action is the carbamates that also vary very much in their toxicity. The most toxic examples, including aldicarb and carbofuran, were only allowed registration in the UK as granules, applied directly into soil, and not as sprays. The less-toxic carbaryl has been very widely used as a broad-spectrum insecticide. Newer groups are the pyrethroids, neonicotinoids and ryanodine insecticides.

Natural pyrethrins had been known for centuries as a potent insecticide, but they were rapidly inactivated, when exposed to sunlight. Research at Rothamsted in the UK led to the development of the synthetic photostable pyrethroids, permethrin, cypermethrin and deltamethrin (Elliott et al., 1973, 1978). Other pyrethroids have been developed, so this group became a very popular broad-spectrum insecticide group.

Similarly, but following the botanical insecticide nicotine, the neonicotinoids, notably imidacloprid, have been developed and rapidly accepted, especially where insects are resistant to the earlier types of insecticide. There are four groups of neonicotinoid insecticide, based on their chemistry: chloropyridyl (e.g. imidacloprid), thiazolyl (e.g. thiamethoxam), furanyl (e.g. dinotefuran) and sulphoximine (e.g. sulfoxaflor).

Neonicotinoids are active at extremely low dosages and have been used as seed treatments as they are absorbed by the plants which are protected over a long period. However, their use in Europe has been limited due to concerns that very low doses are detrimental to bees. In contrast, the authorities in Australia allowed continued use of the neonicotinoids as pollinator declines reported in some parts of the world were likely to be caused by multiple interacting pressures including habitat loss and disappearance of floral resources, honeybee nutrition, climate change, bee pests and pathogens, and miticides and other chemicals intentionally in hives to control varroa mites (Anon, 2014a). However, more favourable weather during the winter of 2013–2014 allowed overwintering bee mortality assessed in 31 countries to drop below 10%, despite neonicotinoid-based crop protection products being still in common use throughout Europe. In the USA, the *Center for Regulatory Effectiveness* (Anon, 2014b) consulted widely and reported that *Varroa destructor* mites are, by far, the greatest threat to feral and managed bees around the world, and secondly that neonicotinoids pose little or no threat to pollinators when used in accordance with regulatory requirements. They also concluded that studies which blame neonicotinoids for contributing to bee health decline are poorly designed and rely on massively overdosing sample bee populations, indicating that the main problems for bee keepers were not the use of the neonicotinoids. This was later confirmed by a study over 3 years to assess chronic sublethal effects on whole honey bee colonies-fed supplemental pollen diet containing imidacloprid at 5, 20 and 100 μg/kg over multiple brood cycles, which showed that field doses relevant for seed-treated crops (5 μg/kg) had negligible effects on colony health (Dively et al., 2015).

A new systemic insecticide flupyradifurone from the butenolide chemical class is effective against sucking pests and could replace imidacloprid, although it has a similar mode of action, but it has a different chemistry. Its development has followed studies of stemofoline alkaloids from a small group of flowering plants (*Stemonacea* family) native to various regions of Southeast Asia, as herbal extracts from these plants have been used for centuries as pesticides and to treat respiratory diseases.

Other new insecticides effective at a low dosages are fipronil, a phenylpyrazole, developed initially as it was extremely effective against locusts, and chlorantraniliprole (Rynaxypyr), very effective against lepidopteran pests. This is the first insecticide from the anthranilic diamide class of chemistry. It affects the ryanodine receptor and muscle function is disrupted, so insects stop feeding, as with the botanical insecticide, ryania. It is an important development as it is effective against insects resistant to older products with a different mode of action.

In contrast to the nerve poisons, insect growth regulators, such as diflubenzuron, affect insect development, mostly by adversely affecting chitin synthesis so the insect fails to complete a moult from one larval stage to the next. Another novel, insecticide, tebufenozide causes larvae to form precocious adults; that is, they attempt to moult into an adult before sufficient larval development has taken place.

Spinosad was the first insecticide using spinosyns A and D. A second-generation insecticide spinetoram, with a mixture of chemically modified spinosyns J and L, has been developed from a fermentation process in which *Saccharopolyspora spinosa* colonies were grown using natural materials such as soybean and cottonseed meal as feedstocks. Another insecticide chlorfenapyr, derived from microbially produced compounds known as halogenated pyrroles, is metabolised into active substance after it is in the pest. It is therefore slow acting compared with other insecticides.

There is now considerable interest in the development of natural organisms such as the fungus *Metarhizium acridum* as a biopesticide, which is very effective against locusts and other acridids. One advantage of mycoinsecticides is that they are selective, but this presents difficulties in marketing a product that is effective against a limited number of pests. Entomopathogenic nematodes have also increased in importance to control certain pests that attack plant roots such as the vine weevil. Biopesticides also tend to be slower acting, but they do integrate well with other biological control agents.

Herbicides

Herbicides are the most extensively used group of pesticides. Already their use is a crucial part of mechanised farming in North America, Europe and Australia. Specific herbicides, such as glyphosate, are linked to genetically modified herbicide-tolerant crops. This has led to their excessive use and increasing problems due to weeds becoming resistant to the herbicide. Herbicides have not been adopted on many small farms which have relied on manual weeding, but the situation is changing as the number of people available for weeding has declined with migration to cities and where disease such as HIV / Aids has increased problems of coping with weed control at the critical stages of early crop establishment. There is still the problem of whether to apply a herbicide if rainfall is erratic and crop establishment uncertain.

Herbicides can act on *contact* with a plant or are *translocated* within the plant. Good spray coverage is needed with contact herbicides. Sometimes only part of the foliage is affected, so some weeds although adversely affected, will survive. Translocated herbicides are particularly important for controlling perennial weeds, such as some of the key grass weeds. An example of a translocated herbicide is glyphosate, which will move down into the rhizomes of grasses, rather than only affect the foliage above ground. As the herbicide is distributed within the plant, good coverage is slightly less important.

Herbicides can also be classified according to the time of application. Weed control may be by means of a *pre-planting* application. This is usually a soil treatment that affects weed seeds before the crop is sown. After the crop has been sown, a *pre-emergence* herbicide will selectively affect the weed species without interfering with the germination and growth of the crop. When farmers have to contend with erratic rainfall and are not sure if a crop can be established, they may opt for a *post-emergence* herbicide applied later to the weeds. The herbicide may be applied to the whole of the crop area, or in the case of post-emergence herbicides, the spray can be applied as a band in the inter-row, or in some cases along the intra-row, using mechanical cultivation of the inter-row. This method is useful with crops that have been genetically modified to be resistant to particular herbicides that can be sprayed over the crop. In contrast, where a crop may be very sensitive to the herbicide, sprays need to be directed to avoid contact with the crop. Individual clumps of weeds can be spot treated, or if certain weeds are confined to specific areas of a field, the farmer can do patch spraying.

Herbicides may be broad spectrum, affecting all types of weeds, or they may be selective. In most cases, selectivity is between monocotyledon weeds, for example grasses and dicotyledons, the broadleaved plants. There are many different groups of herbicides, based on their chemical structure.

The Weed Science Society of America has provided a classification of herbicides (Mallory-Smith and Retzinger, 2003). Most have a very low mammalian toxicity. Most concern of human toxicity has been directed at paraquat, as it is lethal if the concentrate reaches the lungs.

Many different types of herbicides are now available. The following notes refer only to a selected number of different chemical groups.

Amide. Flucarbazone sodium is a relatively new post-emergence herbicide for application on wheat.

Aryloxyphenoxy propionates. These have good activity against grass weeds in broadleaved crops as a post-emergent translocated herbicide. One example is fluazifop-butyl.

Benzoylcyclohexanedione. Mesotrione is a synthetic analog of leptospermone developed to mimic the effects of this natural phytotoxin, obtained from the Californian bottlebrush plant. It is a selective pre- and post-emergence herbicide effective against a range of broadleaved and grass weeds.

Bipyridyliums. Paraquat is the most important in this group. It damages foliage quickly on contact, but is ineffective once the herbicide reaches the soil as it very strongly adsorbed on soil particles. The rapid wilting and desiccation of foliage within hours has enabled effective weed control to be achieved in many crops, where the spray is directed away from the actual crop. It has been extensively used in tree crops such as rubber plantations.

Dinitroanilines. Trifluralin is used as pre-planting soil-incorporated herbicide to reduce the impact of grass weeds in a broadleaved crop. Low water solubility minimises leaching and movement within the soil, but being volatile they must be covered by the soil. Due to its volatility and toxicity to fish and other aquatic organisms, its use is now banned in Europe.

Phenoxy or 'hormone' herbicides such as 2,4-D and MCPA are highly selective for broadleaved weeds, being translocated throughout the plant, affecting cellular division.

Phosphono amino acids, such as glyphosate and glufosinate, are foliar-applied translocated herbicides that interfere with normal plant amino acid synthesis. They are non-selective, but more effective against grasses than broadleaved weeds. There is no soil activity. They are formulated to improve uptake by the plants as rainfall shortly after application can reduce effectiveness.

Pyrazole herbicides (including benzoylpyrazole and phenylpyrazole). This is a new class of herbicides and includes pyroxasulfone, a pre- and post-emergence herbicide with long residual activity. Topramezone is used as a post emergence herbicide. The main target is annual grasses and broad leaves in corn crops, but is recommended for a range of other crops.

Substituted ureas. Most of these, such as isoproturon, flumeturon, diuron and linuron are non-selective, pre-emergence herbicides, which are absorbed in the soil and then taken up by roots. Some are active as foliar-applied post-emergence herbicides. Isoproturon has been widely used, especially to control black grass in winter wheat crops, but its extensive use had led black grass becoming resistant to it. However concern was raised about a risk to aquatic organisms by its movement into surface water courses due 'run-off' and via land drains. No risk management methods could be identified and its registration was withdrawn in the UK.

Sulfonylureas. This is a large group that is used mainly to control broad-leaved weeds by inhibiting meristematic growth. Metsulfuron-methyl. and others in the group have both foliar and soil activity and are active at extremely low application rates – a few grams per hectare. Rimsulfuran has been used to control glyphosate-resistant weeds such as rye grass. It may be used pre-emergence in maize, but with a safener (e.g. isoxadifen) has been used post-emergence. If small amounts of these herbicides remain in the soil too long, the following crop may be affected.

Triazines. This group includes one of the most commonly used herbicides, atrazine, which was very effective as a post-emergence spray in maize. However, it has been implicated in environmental problems, as it has been claimed that very low doses in water have an endocrine disruption effect that has resulted in a decline in frog populations, so its use has been curtailed.

Fungicides

The use of sulphur to protect vines dates back to ancient Greek civilisations, and with Bordeaux mixture since the end of the nineteenth century, most developments of fungicides have occurred only in the last few decades. Apart from the contact, protectant fungicides, such as copper fungicides and mancozeb, a number of systemic fungicides (Table 1.2) with different modes of action have been developed, most recently the stobilurins. Unfortunately, pathogens that are susceptible to a particular type of fungicide often become less sensitive. Thus great care is needed to avoid selection of pathogens resistant to a fungicide, by only applying those with a particular mode of action for a short period before using another one with a different mode of action in rotation. Manufacturers have also recommended mixtures as a means of delaying selection of resistant strains. Fungicidal seed treatments are important to protect young seedlings. Among the new fungicides isopyrazam, a pyrazole carboxamide is used to control black sigatoka disease on bananas. Benzovindiflupyr is being introduced to control Asian soybean rust (*Phakopsora pachyrhizi*).

Oliver and Hewitt (2014) provide an updates on the emergence of the strobilurins and succinate-dehydrogenase inhibitors (SDHIs) and the

Table 1.2 Some examples of fungicides

Type of fungicide	Example
Triazoles	propiconazole metconazole tebuconazole
Benzenoid	mefenoxam
Morpholines	fenpropimorph
Anilinopyrimidines	cyprodinil, pyrimethanil
Benzimidazoles	carbendazim
Carboxamides	isopyrazam
SDHI	boscalid fluxapyroxad penthiopyrad
Strobilurins	azoxystrobin
Ethylbenzamide	fluopyram (also marketed as a nematicide)
Host Plant Defence Induction; Group P1	acibenzolar-S-methyl
Others	chlorothanil

increased incidence of fungicide resistance. They also discuss legislative requirements to reduce fungicide applications in IPM programmes to minimise selection of resistance.

Rodenticides

Significant crop losses can be caused by rodents, both in the field and in stores. Rats are also a major problem in cities and other areas where they can get food. Various poisons have been set out in baits, usually inside traps to prevent other mammals, especially dogs from gaining access to the poison. Following the use of the anti-coagulant warfarin, to which rats have become resistant, other rodenticides such as bromadiolone and difenacoum have been introduced. There is particular concern that predatory birds can be affected by eating rodents that have consumed a poisoned bait, but have not yet died.

Crop distribution

The distribution of pesticide use is illustrated in Figures 1.2–1.4. Public concern is directed mainly at the amounts of insecticides and fungicides used on food crops, especially those that are eaten without further processing. IPM was developed to reduce pesticide use by encouraging farmers to combine different control techniques. IPM is now adopted as requirement in Europe aimed at using pesticides only as a last resort. While consumers may prefer to have produce without any pesticide residue, the lower yields

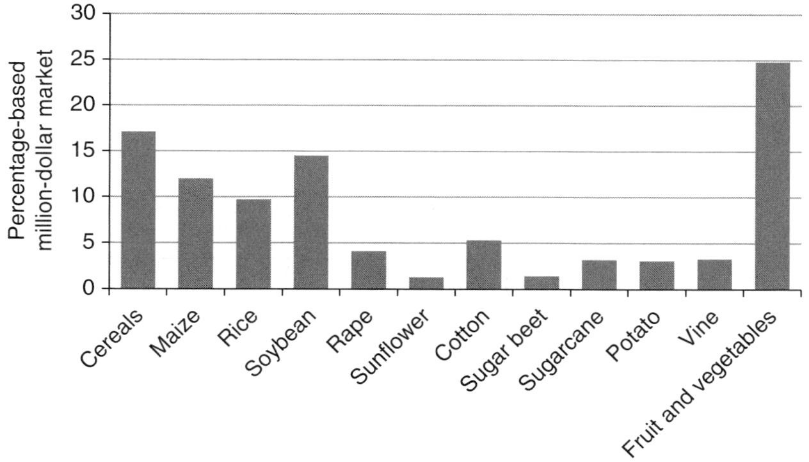

Fig. 1.4 Global sales of pesticides by major crops (Phillips McDougall, 2018).

and cost of 'organic' crop production indicates that some pesticides will be needed for rapid action against pests affecting yield and quality of produce. Where GM crops have been accepted, those with a Bt gene are grown can reduce insecticide use against some pests. In contrast the development of herbicide-resistant crops will see an expansion of the use of certain herbicides. Newer, less-toxic pesticides, including biopesticides are being developed and will also be crucial in maintaining high yields of crops.

Major crops

The application of pesticides has been an important component of changes in agricultural practices, including new crop varieties that have enabled yields of major crops to be increased. While they have not increased the yield potential, they have enabled farmers to realise a higher proportion of the potential yields by reducing the losses due to pests and pathogens and from weed competition. In addition improved quality of the harvested produce has allowed longer storage under suitable conditions that enables marketing of the crop to be extended. A few examples of the higher yields harvested are shown for the following selected crops.

Wheat

Yields of wheat worldwide average only 2.6 t/ha, although the potential is much higher as shown by the improvement in yields achieved in the UK (Table 1.3), although yields have reached a plateau. For higher yields,

Table 1.3 Area and yield of wheat in the UK

Year	ca. 1932	1969–1971	1971–1981	1988–1990	1999	2003	2013
Area harvested (1000 ha)		980	1434	1994	1847	1837	1510
Yield (mt/ha)	2.1	4.2	5.6	8.8	8.0	7.8	7.8

Table 1.4 Yields of rice (t/ha rough rice) from http://www.irri.org/science/ricestat/pdfs and http://ricestat.irri.org:8080/wrs2/entrypoint.htm

Year	Global	China	India	Japan
1962	1.89	2.08	1.54	5.14
1972	2.32	3.25	1.60	5.85
1982	2.98	4.89	1.85	5.69
1992	3.59	5.90	2.61	6.28
2002	3.92	6.27	2.91	6.63
2012	4.40	6.74	3.52	6.74

research has been initiated to increase the potential yield from 8 to 20 t/ha with new varieties. Much of the yield benefit in the UK has been due to efficient weed management following the introduction of herbicides. A return to the days of manual weeding is unthinkable as the cost of labour would be too high. In the UK, with organic agriculture, the estimate for casual labour for some vegetable crops can be as much as 40 days per hectare at £5.70 per hour, although mechanical hoeing would be done where possible to avoid manual weeding. In the USA, a state law was enacted to ban weeding crops with short-handled hoes as the work was excessively arduous, but the use of long-handled hoes was considered to cause some crop damage.

Rice

Success with breeding new high-yielding varieties of the 'Green Revolution' in Asia led to higher yields and production (Table 1.4), but also increased pest problems. The use of insecticides is generally blamed for the outbreaks of the brown planthopper, *Nilapavata lugens*, as insecticides were promoted in some areas as if they were like fertilizers and increased yields. In practice, poorly applied broad-spectrum insecticides made the planthopper problem worse as little spray reached the lower part of the stem favoured by the nymphs. The pest problem was also due to the overlapping of two or more rice crops with little attention given to a closed season between harvesting and sowing a second crop. Improvements in variety selection, enabling farmers to sow resistant varieties reduced the planthopper problem and by avoiding any insecticide use in the first six weeks of plant development, natural enemies have been able to exert adequate control of most pests (Way and Heong, 1994). Farmer field schools have been more effective in lowland irrigated rice areas as the system was based on extensive research at the International Rice Research Institute (Matteson, 2000). One of the problems in adopting IPM is getting farmers to accept that crop losses are not always as high as they perceive (Escalada and Heong, 2004).

However, rice farmers also have to contend with weeds as the increased cost of labour has resulted in changes from the transplanting of seedlings to more extensive use of direct seeding. Yield losses as high as 46% caused by weeds have been reported, so in some areas, farmer adoption of herbicides has increased rapidly in the last decade, although alternative crop establishment methods have also been adopted to reduce weed problems. Crops may need to be sprayed with fungicide in some areas due to diseases, such as rice blast.

Cotton

Insecticide use on cotton was considerable as insect pests have been a major constraint on production. Before the discovery of DDT, yields were generally less than 500 kg seed cotton per hectare obtained on 'organic cotton' but even small-scale farmers in Africa could expect to get yields of over 1000 kg/ha (Tunstall and Matthews, 1966; Gower and Matthews, 1971). This has changed with the development of GM cotton incorporating genes encoding toxin crystals in the Cry group of endotoxin of *Bacillus thuringiensis*. The *Cry1Ac* gene was used initially and its effectiveness has been improved by the addition of the *Cry2Ab* gene. The success of this development is due to

Fig. 1.5 Contrast between untreated cotton with many insects and sprayed crop ready for harvesting in Malawi. Similar effect now between Bt cotton and untreated plants with severe bollworm infestation (Photograph by Graham Matthews).

the first-instar bollworm larvae contacting the toxin as soon as they start to feed. This enables growers to achieve high yields without multiple sprays during a season to control the bollworms and other lepidopteran pests. The toxin is not effective against sucking pests, such as aphids, so some sprays may be required unless in the absence of the bollworm spray programme, natural enemies will control sucking pests. Where Bt cotton is not grown, farmers have to continue to monitor their crop and spray when necessary according to action thresholds. As with other insecticides, prolonged use of Bt cotton will result in a key pest becoming resistant to the Bt toxin. In contrast to the USA where a resistance management policy was implemented, resistance of the pink bollworm (*Pectinophora gossypiella*) to Bt cotton has been recorded in India (Fabrick et al., 2014).

Maize

Pesticide use on maize is also changing with GM Bt-maize incorporating the *Cry1Ab* gene to give resistance to corn root worm and corn earworm. Weed management is simplified with herbicide-tolerant maize. Yields of over 9 t/ha are achieved, whereas farmers in many areas of Africa barely produce 0.5 t/ha. Their major problem is initially weeds, which are highly competitive with young seedlings during the first 3 weeks after seed germination. In parts of Africa, the parasitic weed *Striga* has continued to be a major problem with crop losses over 50% under moderate to severe infestations (Parker,

Fig. 1.6 Manual weeding of maize. Greater use of herbicides is now likely in areas where labour is not so readily available (Photograph by Graham Matthews).

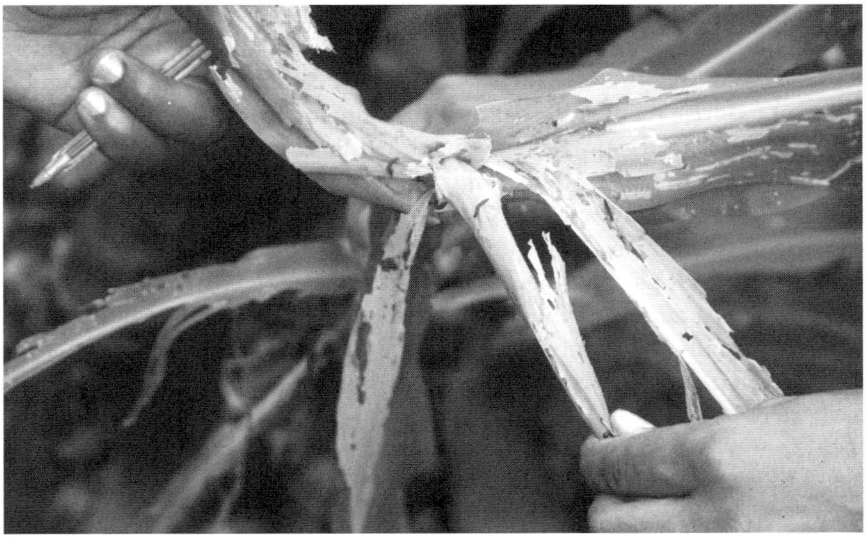

Fig. 1.7 Maize damaged by stem borer (Photograph by Graham Matthews).

Fig. 1.8 Simple granule treatment where Bt maize is not yet grown (Photograph by Graham Matthews).

2012). Sowing seeds of a maize-tolerant variety treated with imidazolinone herbicide can be very effective, when soil moisture is suitable (De Groot et al., 2008, Kabambe et al., 2008). Progress has also been made with new varieties resistant to *Striga* by the International Institute for Tropical Agriculture.

Traditionally African farmers have hoed their crops, but the amount of time and effort needed often results in part of the sown area being abandoned. Attention is now being directed in Africa to conservation farming which will introduce more herbicide usage. Locusts can also decimate young maize crops, but these can be controlled using a biopesticide – *Metarhizium acridum*. Without GM maize, stem borers can be controlled by a relative small amount of insecticide, provided it is in the whorl of leaves of the young plants. Crop protection is again crucial when the grain is harvested.

Fruit

Bananas

A major disease of the 'Cavendish' variety (about 10% of global production) of bananas, Sigotoka, has resulted in growers resorting to fungicide applications, usually applied by aircraft on large estates. A new form of the disease, black sigatoka and a new strain of fusarium wilt, also known as Panama disease, is causing particular concern, as this disease is far less easy to control. Bananas, with plantains, are widely grown on small farms in Africa, largely for local consumption, but a major drop in production in Uganda occurred due to failure of disease control in 1980, plus damage due to nematodes and banana weevils. Some farmers are now growing new resistant varieties imported from Central America. Meanwhile a major effort is underway to develop new disease resistant varieties.

Apples

Apart from a number of insect pests, such as the codling moth, apple orchards can suffer from mildew and scab diseases. For insect control, where insecticides are used the trend has been to apply insect growth regulators such as pyriproxyfen and tebufenozide, although indoxacarb and spinosad are also recommended. Much emphasis has been put on controlling insects with pheromone traps and encouraging natural enemies, but several fungicides sprays may be needed during the season. In the UK research is aimed at endeavouring to control the pathogen late in the season after harvesting to reduce the carryover of infection to the following season. Fewer early season fungicide sprays should then control the disease and also reduce the likelihood of any pesticide residues in the apples.

Fig. 1.9 Cocoa farmers need apply fungicides to protect pods from black pod disease (Photograph by Roy Bateman). (*See insert for color representation of the figure.*)

Vegetables

Potatoes

Commercial yields of potatoes vary from around 18 to over 45 t/ha depending on the variety and soil type, but also on protection from nematodes, late blight and insect pests. In the UK, the average yield is about 45 t/ha. If untreated, late blight, which was the cause of the Irish famine (1846–1850), can spread very rapidly with as much as 75% of foliage destroyed in less than 10 days. In fungicide trials yield increases of up to 30 t/ha have been reported. A GM potato with blight resistance has been developed but not yet commercialised. Similar devastating crop damage can also be inflicted by the Colorado beetle (*Leptinotarsa decemlineata*), which has spread from the USA across Europe to Asia. Yield loss due to viruses transmitted by aphids is usually low in the year in which the crop acquires infection, but if those tubers are used as seed potatoes, the yield will decline rapidly. Thus farmers obtain certified seed potatoes from areas with low aphid infestations. However, aphids may still need to be controlled if populations build up rapidly. Ideally crop rotation is used to minimise nematode damage, but nematicides are still required where potatoes are grown one year in four on the same land.

Tomatoes

In the tropics tomatoes are grown in fields, but in Mediterranean and temperate climates the crop is in plastic or glasshouses. Yields as high as 200 t/ha have been harvested, but protection from pests and diseases is essential. In a more controlled environment, the trend has been away from using insecticides to greater reliance on biological control, but protection from several diseases is still essential.

Forests

Certain insect pests can cause major defoliation of large areas of forests. In North America the spruce budworm (*Choristoneura fumiferana*) and the gypsy moth (*Lymantria dispar*) are among the key pests that have led authorities to spray large areas with insecticides. In the early days of these programmes broad-spectrum insecticides were used, but currently *Bacillus thuringiensis* and other more ecological acceptable products are sprayed. In Poland control of the nun moth (*Lymantria monacha*) was achieved over 2.5 million hectares using aerially applied Bt and the chitin inhibitor, diflubenzuron in 1994–1997. Thus control operations are crucial in some years to preserve forests.

Tillage

Farmers have for centuries used crop rotation and traditional tillage by ploughing and hoeing to manage weeds in the fields. However, in some parts of the world, ploughing may adversely affect earthworms and the loosening of the soil makes it prone to erosion. There is, therefore, awareness that for some crops reducing tillage, usually referred to as conservation tillage, has advantages. The aim is to protect the soil from the damaging effects of rain splash by leaving 30–50% stover on the soil surface to retain more rain on the fields. Various techniques have been developed to sow and plant the crop, for example by using a narrow furrow or just individual planting holes. Most of the land is undisturbed, but with the lack of burying weed seeds, conservation tillage does depend on careful use of herbicides to avoid weed competition.

Amenity areas and home gardens

Significant quantities of pesticides are now used by local authorities, for example in keeping pathways and gutters free of weeds. More emphasis has been given to non-chemical methods where these are effective, but cost-effective treatments are required. A limited number of pesticides are now marketed for home and garden use, notably for controlling insect pests on tomatoes, roses and lawns. In the UK, only those considered to be safe to use without professional training and do not require protective clothing are permitted. Many of these products have been sold in ready to use in small plastic containers incorporating a trigger-operated nozzle. In some countries the use of pesticides on turf and in gardens has been prohibited by local bye-laws, but in these areas where serious pest damage occurs, professional pest control companies may be employed.

Nuisance pests and vector control

The control of ants and cockroaches in dwellings was often done by applying a low concentration dust, to areas where the insects are known to live. The alternative has been to use pressure packs, known more usually as aerosol cans. Professional operators controlling nuisance pests in restaurants, hospitals, aircraft and other locations use a range of different spray equipment.

In more tropical climates vectors of malaria, dengue and other diseases need to be controlled. Insecticide-treated bed nets have become widely available and reduce illness and mortality of those who remain under the bed net while the vectors are active. Indoor residual spraying has also given good control of mosquitoes that enter houses, but in some situations an area-wide space treatment is needed. Urban areas can be treated with vehicle-mounted cold fogging equipment or aircraft may be used to apply insecticide at the flight time of the mosquitoes. Mosquito control units are present in most counties throughout the USA and these have been particularly active following the outbreak of West Nile Virus. Reduction in larval breeding sites by drainage may require follow up treatment of areas without drainage by applying larvicides.

Legislation

Legislation on the registration and use of pesticides has been primarily by national governments. In Europe Directive 91/414/EEC initiated harmonisation of the registration procures, and has been followed by the Sustainable Use Directive (SUD) new legislation in the form of a Regulation (EC Regulation 1107/2009), which came into force in June 2011 and requires compliance by all EU countries. The aim has been to minimise risks of environmental pollution based on data obtained from manufacturers and to exclude the most hazardous compounds. This is important as other parts of the world, especially in areas where protective clothing is either not available or considered too uncomfortable to wear, may follow and no longer register many highly toxic pesticides. Apart from registration of the pesticides, the regulation also brings in requirements for pesticide packaging with more emphasis on recycling of cleaned pesticide containers to increase safety. The number of pesticides that can be marketed in Europe has already been significantly reduced as a result of this legislation and has also affected countries exporting crops to Europe as these must also comply with regulations on maximum residue levels (MRLs).

At the same time, the Machinery Directive 2006/42/EC, has been amended and requires new pesticide application equipment to meet set standards prior to being marketed. Equipment now requires regular checks and spray operators must receive training to improve the precision of pesticide application, both in terms of placement and when an application is needed to minimise the amount of pesticide used in the environment. In Europe each country must develop national plans which promote the use of alternative controls with the application of any pesticide regarded as a last resort as part of IPM.

Requirements for registration are discussed in Chapter 2. In the international sphere, the Food and Agriculture Organisation (FAO) has sought to achieve harmonisation of the data requirements since the *Ad Hoc* Government Consultation on Pesticides in Agriculture and Public Health (FAO, 1975). Following this meeting, FAO published a Code of Conduct on the Distribution and Use of Pesticides, which has been subsequently amended and is now the FAO Code of Conduct on Pesticide Management (FAO, 2012). It includes requirements for Prior Informed Consent (PIC) aimed at assisting the less developed countries without the resources to administer a full registration system to decide whether it should allow the import of certain pesticides. FAO has also published a number of guidelines in support of the Code (see http://www.fao.org/agriculture/crops/thematic-sitemap/theme/pests/code/list-guide-new/en/). Under the Rotterdam Convention, exporters of pesticides have to inform importers in developing countries about the toxicity and hazards associated with the use of products included on the PIC list and receive their authority before the products can be exported.

The FAO Code is voluntary, but the requirements for PIC have been included in a European Community regulation applicable by law in the Member States. A PIC database is maintained at the International Register of Potentially Toxic Chemicals held at Geneva, where the International Programme on Chemical Safety (IPCS) is located. Under the Stockholm Convention a number of pesticides are included in the list of persistent organic pollutants (POPs) and their use is now banned, although there is a derogation for DDT used only for indoor residual spraying to control mosquitoes.

WHO has published a classification system (WHO, 2010) (Table 1.5) for pesticides, based on the acute toxicity of the formulation. Class I pesticides are the most hazardous to use, whereas those in the unclassified category are the least toxic to mammals. The examples in Table 1.6 show that Class I are mostly the older types on insecticides. The Codex Alimentarius Commission of the United Nations is responsible for harmonisation of

Table 1.5 WHO classification (http://www.who.int/ipcs/publications/en/pesticides_hazard.pdf)

		Oral toxicity[#]		Dermal toxicity[#]	
Class	Hazard level	Solids*	Liquids*	Solids*	Liquids*
Ia	Extremely hazardous	<5	<20	<10	<40
Ib	Highly hazardous	5–50	20–200	10–100	40–400
II	Moderately hazardous	50–500	200–2000	100–1000	400–4000
III	Slightly hazardous	>500	>2000	>1000	>4000
U	Unclassified				

[#]Based on LD_{50} for the rat (mg/kg body weight).
*The terms 'solids' and 'liquids' refer to the physical state of the product or formulation being classified.

standards related to the international food trade and by collaboration with the Joint meetings of a FAO Working Party and a WHO Expert Committee, the Codex Committee on Pesticide Residues sets international standards. The International Union of Pure and Applied Chemistry (IUPAC) also has a role in setting specifications for each pesticide.

Some countries still do not have adequate legislation or trained staff to register pesticides and ensure that only registered pesticides are available to farmers. This lack of regulation is of great concern due to the risk of human health problems, associated with the most toxic insecticides being used by illiterate farmers, usually without training or adequate protection during their application (Matthews et al., 2011; van den Berg et al., 2011).

Estimates of poisoning cases are not easy in many countries with a poor infrastructure. In some countries, many affected by poisoning may not see a doctor and only a small proportion reach a hospital for proper treatment. In 2006, the WHO estimated global pesticide poisoning at 3 million cases, although the estimate may not represent the all cases of illness and death due to misdiagnosis or non-hospitalized cases (Mancini et al., 2009). In developed countries Thundiyil et al. (2008) indicated a rate of acute pesticide poisoning among agricultural workers was 18.2/10,000 full time workers and 7.4 per million among schoolchildren.

The most horrific number of deaths was at Bhopal in India in 1984, when a chemical methyl isocyanate (MIC) used at a factory making the carbamate insecticide carbaryl (Sevin) was contaminated with water. The reaction led to an extremely toxic gas escaping and this killed nearly four thousand people in the following hours. Regrettably, vast numbers had been allowed to live in slums close to the factory and with no contingency plans; these people were not protected from the toxic gas. Even more, further away from the factory were affected by the gas and suffered severe health problems. Some estimates indicate as many as 15,000 died later, with many more continuing to suffer from chronic symptoms. In 2006, the government confirmed that the leak had caused 558,125 injuries including 38,478 temporary partial injuries and approximately 3,900 severely and permanently disabling

Table 1.6 Examples of pesticides according to WHO classification in relation to mammalian toxicity for the active substance. The type and concentration of the formulation will adjust the ranking, thus pyrethroid insecticides are used at a low concentration, and are considered less hazardous

Class	Insecticide	Fungicide	Herbicide	Rodenticide
Ia	aldicarb	captafol		brodifacoum
	mevinphos			
	parathion			
	phorate			
	phosphamidon			
Ib	azinphosmethyl			warfarin
	carbofuran			
	dichlovos			
	formetanate			
	methamidophos			
	methomyl			
	monocrotophos			
	nicotine			
	triazophos			
II	bendiocarb	azaconazole	2,4-D	
	carbosulfan	copper sulphate	paraquat	
	chlorpyrifos	fentin hydroxide		
	cypermethrn	tetraconazole		
	deltamethrin			
	dimethoate			
	fenitrothion			
	fenthion			
	fipronil			
	imidacloprid			
	lambda-cyhalothrin			
	rotenone			
	thiodicarb			
III	acephate	copper hydroxide	ametryn	
	amitraz	copper oxychloride	bentazone	
	malathion	metalaxyl	dicamba	
	spinosad	thiram	dichlorprop	
	spirotetramat			
	resmethrin		glufosinate	
	trichlorfon		isoproturon	
			linuron	
			MCPA	
			mecoprop	
			propanil	
Unclassified	phenothrin	axoxystrobin	atrazine	
	spinetoram	benomyl	pyraxosulfone	
	temephos	carbendazim	simazine	
		iprodione	trifluralin	
		mancozeb		
		sulphur		

(a)

(b)

Fig. 1.10 Contrasts in protective clothing while using a lever-operated knapsack sprayer (a) India (Photograph by Graham Matthews), (b) Pakistan (Photograph by Graham Matthews), (c) the UK (Photograph courtesy of Hardi International) and (d) manually carried lance in southern Europe (Photograph by Richard Glass).

(c)

(d)

Fig. 1.10 (*Continued*)

injuries. Subsequently an Indian court sentenced seven ex-employees in 2010 to 2 years imprisonment and a fine of about $2,000 each, for causing death by negligence.

In the Soviet Union, the use of pesticides had expanded so much that by the 1980s the USSR was one of the world leaders in pesticide use in terms of per hectare and per capita. Unfortunately their pesticides were often of inferior quality, packaged in large containers, poorly stored and inefficiently applied, often by aircraft. This led to vast numbers of people being poisoned, with for example the average daily concentration of OPs such as demeton over 0.1 mg/m^3 in air, 500–1000 m from cotton fields (Fedorov and Yablokov, 2004). This extensive poisoning in Uzbekistan led to a switch to biological control with the setting up of biofactories to produce parasitoids, a practice still adopted in the country.

Globally, the cause and symptoms of poisoning vary between chemicals and countries (Harris, 2000). In some cases, where deaths have occurred, it is undoubtedly due to application of pesticides classified by WHO as Ia or Ib pesticides with no protective clothing being worn. Some of these organo-phosphorus insecticides, such as parathion, methamidophos and monocrotophos, were used because farmers perceived that these killed their pests quickly. In some countries deliberate drinking of pesticides in suicide attempts has been the main cause of death, rather than occupational exposure.

In Sri Lanka, the total national number of admissions due to poisoning doubled between 1986 and 2000, with an over 50% increase in admissions due to pesticide poisoning, but the number of deaths fell. In particular the number of deaths due to the organo-phosphate insecticides, monocrotophos and methamidophos fell from 72% of pesticide-induced deaths as the import was restricted and eventually banned in 1995. However the use of these insecticides was replaced by endosulfan (WHO Class II) and this led to a rise in deaths from 1 in 1994 to 50 in 1998 when this insecticide was also banned. Over the decade the number of deaths due to pesticide poisoning had not changed significantly with WHO Class II OP insecticides becoming a major factor. The switching from one pesticide to another, especially in relation to self-poisoning needs further attention and although legislation on pesticides had an effect, the emphasis now must be on other strategies to reduce the availability of the most hazardous chemicals (Konradsen et al., 2003; Roberts et al., 2003). One strategy is for pesticides to be kept in special locked containers to reduce access to the poisons (Gunnell et al., 2007).

In South Africa early indications are that small-holders, who adopted the growing of Bt cotton had a reduced incidence of skin disorders, feeling generally unwell and other health effects that had been associated with spraying for bollworm. It has been suggested that if all farmers grew Bt cotton, the number of poisonings would decrease to just two per season, compared to 51 reported cases in the 1997/1998 season (Bennett, 2003). Similar reports come from China (Hossain et al., 2004).

Various non-governmental organisations (NGOs) have lobbied for pesticide reduction policies, including the PAN, which has sought to eliminate the hazards of pesticides, reduce dependence on them and prevent unnecessary expansion of their use, while increasing sustainable and ecological alternatives to chemical pest control.

Unfortunately in contrast to demands for banning of many pesticides, less attention has been given to equipment leaving it to farmers to choose and maintain their sprayers. In consequence, often cheap and poorly maintained sprayers are used and this has frequently resulted in prolonged exposure to pesticides, especially by those who have poor facilities to wash after work. This is likely the cause of many cases of poisoning, especially with insecticide sprays. In order to prevent leakage of pesticide from the sprayer tank over the operator's body or leakage over unprotected hands, FAO has published minimum standards for pesticide application equipment (Anon, 2001).

Compared to tropical areas, relatively few cases of acute poisoning are reported in temperate climates, where protective clothing is available and worn. In the UK, the extremely hazardous pesticides, such as aldicarb, applied as solid granules to soil for nematode control have been withdrawn, together with other pesticides, either because the commercial company did not consider sales justified the cost of the additional test data needed to meet the current requirements of the EU, or additional data and evaluation has led to revocation of registration.

The National Poisons Information Service (NPIS) in the UK has identified cases that required health care (Perry et al., 2014) and shown that from 2004 to 2013, 7,804 cases of pesticide exposure were identified out of 34,092 enquiries. Eighty-seven per cent were unintentional acute and 9.7% as acute, deliberate self-harm, while the remainder were unintentional but chronic exposures. Of the 38 deaths recorded, most were due to either paraquat or diquat (14) or aluminium phosphide (12). Most of the unintentional exposures were due to children having access to bait-type products for rodent or ant control. NPIS found that the minimum incidence of pesticide exposure requiring health care contact was 2.0 cases/100,000 population per year. This was higher than that reported by other toxicovigilance schemes such, the Pesticides Incidents Appraisal Panel (PIAP), essentially covering agricultural use, which has reported a decrease in the number of acute 'health incidents' per annum.

Another concern has been the presence of pesticide residues in food. Regulatory authorities analyse food samples (see Chapter 7) to determine whether the residues exceed the maximum residue level (MRL) that may occur following good agricultural practice. Approximately 1% of samples of food grown in the UK examined for pesticides residues in 2012 exceeded the maximum residue level compared with 7% for produce imported from outside the EU. In many countries, this concern has led to a policy to reduce the amount of pesticide or number of applications in a season. It is quite easy to reduce the quantity applied when a more active molecule, applied at a few grams per hectare can be applied instead of an older product. Reducing dosage of an application may be possible if it is correctly formulated and applied at the optimum time, but this is not always possible due to weather conditions. However, the policy encouraged the need for research into alternative strategies of pest control and in emphasising a need for integrating different control tactics has led to the mandatory IPM policy.

While some governments considered adding a tax on pesticides to reduce their use, in the UK in response to this threat, the Crop Protection Association introduced a voluntary initiative (VI) aimed at improving the standards of pesticide use through research, training, stewardship and communication (see www.voluntaryinitiative.org.uk/). This initiative has been highly successful with over 20,000 spray operators now registered on the National Register of Sprayer Operators (NRoSo) and over 90% of sprayed areas are treated with equipment that has been tested.

In 1978 a scheme, known as BASIS, was established in the UK by the pesticide industry to develop standards for the safe storage and transport of agricultural and horticultural pesticides and to provide a recognised means of assessing the competence of staff working in the sector. BASIS provides the training for staff in companies marketing pesticides, including those providing products used for amenity areas.

In the UK in 1996, another group, the 'Pesticide Forum' was set up to bring together a wide range of organisations representing those who make, use or advise on pesticides as well as environmental, conservation and consumer interests. The forum continues to provide a mechanism for exchanging ideas and for encouraging joint initiatives to address particular issues. It also provides advice to Government on pesticide usage matters. It reports progress annually on a number of indicators covering economic, environmental and social issues including compliance on water quality (a 30% reduction in the frequency of detection of individual pesticides in untreated surface water at levels above 0.5 and 0.1 ppb), and benefits to biodiversity by adoption of crop protection management, now IPM programmes and changes in the behaviour of farmers through training (see http://www.pesticides.gov.uk/guidance/industries/pesticides/advisory-groups/pesticides-forum/pesticides-forum-annual-reports).

Although many of the foods and beverages we consume contain natural pesticides to defend plants from pest attack, many people have a preference for 'organic' produce, completely free of farmer-applied chemicals. The aim of organic farming is to develop good soil with healthy crops that have natural resistance to pests and diseases, and to use crop rotations to encourage natural predators. However, organic standards allow seven pesticides that are either of natural origin (e.g. soft soap) or simple chemical products – copper compounds and sulphur, which occur naturally in the soil.

This chapter has shown that there have been many changes in pesticide regulation in the last decade in response to concerns about their use. IPM is now a policy within the European Union. Nevertheless, farmers continue to need to apply pesticides to maintain high yields to feed a growing global population. In the following chapters, the way in which governments regulate their use is described and ways in which we can protect people.

References

Andersons Centre (2014) The Effect of the Loss of Plant Protection Products on UK Agriculture and Horticulture and the Wider Economy. The Andersons Centre, Melton Mowbray.

Anon (2001) *Guidelines on Minimum Requirements for Agricultural Pesticide Application Equipment*. FAO, Rome.

Anon (2014a) *Neonicotinoids and the Health of Honey bees in Australia – Overview Report*. Australian Pesticides and Veterinary Medicines Authority, Symonston.

Anon (2014b) *Is Varroa destructor or Neonicotinoid Pesticides Responsible for Bee Health Decline?* Report to the President by the Pollinator Health Task Force. Center for Regulatory Effectiveness. Washington, DC.

Anon and Committee on Toxicity of Chemicals in Food, Consumer Products and the Environment (1999) *Organophosphates*. Department of Health, London.

Bennett, R., Buthelezi, T.J. Ismael, Y. and Morse, S. (2003) Bt cotton, pesticides labour and health: a case study of smallholder farmers in the Makhathini Flats, Republic of South Africa. *Outlook on Agriculture* **32**, 123–128.

van den Berg, H., Hii, J., Soares, A., Mnzava, A., Ameneshewa, B., Dash, A. P., Ejov, M., Tan, S.H., Matthews, G., Yadav, R.S. and Zaim, M. (2011) Status of pesticide management in the practice of vector control: a global survey in countries at risk of malaria or other major vector-borne diseases. *Malaria Journal* **10**, 125.

Carson, R. (1962) *Silent Spring*. Hamish Hamilton, London.

Crauford-Benson, H.J. (1946) Naples typhus epidemic 1943–4. *British Medical Journal* **1**, 579–680.

De Groot, H., Wangare, L., Kanampiu, F., Odendo, M., Diallo, A., Karaya, H. and Friesen, D. (2008) The potential of a herbicide resistant maize technology for *Striga* control in Africa. *Agricultural Systems* **97**, 83–94.

Dively, G.P., Embrey, M.S., Kamel, A., Hawthorne, D.J., Pettis, J.S. (2015) Assessment of chronic sublethal effects of imidacloprid on honey bee colony health. *PLoS One* **10**(3), e0118748.

Elliott, M., Farnham, A.W., Janes, N.F. Needham, P.H. and Pulman, D.A. (1973) Potent pyrethroid insecticides from modified cyclopropane acids. *Nature* **244**, 456.

Elliott, M., Janes, N.F. and Potter, C. (1978) The future of pyrethroids in insect control. *Annual Review of Entomology* **23**, 443–469.

Escalada, M.M. and Heong, K.L. (2004) A participatory exercise for modifying rice farmers' beliefs and practices in stem borer loss assessment. *Crop Protection* **23**, 11–17.

Fabrick, J.A., Ponnuraj, J., Singh, A., Tanwar, R.K. Unnithan, G.C., Yelich, A.J., Li, X., Carriere, Y. and Tabashnik, B.E. (2014) Alternative splicing and highly variable cadherin transcripts associated with field-evolved resistance of pink bollworm to Bt cotton in India. *PLoS One* **9**, e97900.

FAO (2012) FAO Code of Conduct on Pesticide Management. FAO, Rome.

Fedorov, L.A. and Yablokov, A.V. (2004) *Pesticides: The Chemical Weapon that Kills Life: The USSR's Tragic Experience*. Pensoft, Sofia/Moscow.

Food and Agriculture Organisation (FAO) (1975) Report of Ad Hoc Government Consultation on Pesticides in Agriculture and Public Health. FAO, Rome.

Gower, J. and Matthews, G.A. (1971) Cotton development in the southern region of Malawi. *Cotton Growing Review* **48**, 2–18.

Gunnell, D., Fernando, R., Hewagama, M., Priyangika, W.D.D., Konradsen, F. and Eddleston, M. (2007) The impact of pesticide regulations on suicide in Sri Lanka *International Journal of Epidemiology* **36**, 1235–1242.

Harris, J. (2000) *Chemical Pesticide Markets, Health risks and Residues*. Biopesticides Series 1 CABI Publishing, Wallingford.

Hossain, F., Pray, C.E., Lu, Y., Huang, J. and Fan C. (2004) Genetically modified cotton and farmers' health in China. *International Journal of Occupational and Environmental Health* **10**, 296–303.

Kambambe, V.H, Kanampiu, F. and Ngwira, A. (2008) Imazapyr (herbicide) seed dressing increases yield, suppresses *Striga asiatica* and has seed depletion role in maize (*Zea mays* L.) in Malawi. *African Journal of Biotechnology* **7**, 3293–3298.

Karalliedde, L., Feldman, S., Henry, J. and Marrs, T. (eds) (2001) *Organophosphates and Health*. IC Press, London.

Konradsen, F., van der Hoek, W., Cole, D.C., Hutchinson, G., Daisley, H., Singh, S. and Eddleston, M. (2003) Reducing acute poisoning in developing countries: options for restricting the availability of pesticides. *Toxicology* **192**, 249–261.

Lechenet, M., Bretagnolle, V., Bockstaller, C., Boissinot, F., Petit, M-S., Petit, S. and Munier-Jolain, N.M. (2014) Reconciling pesticide reduction with economic and environmental sustainability in arable farming. *PLoS One* **9**(6), e97922.

Lodeman, E.G. (1896) *The Spraying of Plants*. Macmillan, London, 399pp.

MacBean, C. (2012) *The Pesticide Manual*. BCPC Publications, Farnham.

Mallory-Smith, C.A. and Retzinger, E.J. (2003) Revised classification of herbicides by site of action for weed resistance strategies. *Weed Technology* **17**, 605–619.

Mancini, F., Jiggins, J.L.S. and O'Malley, M. (2009) Reducing the incidence of acute pesticide poisoning by educating farmers on integrated pest management in South India. *International Journal of Occupational and Environmental Health* **15**, 143–151.

Matteson, P.C. (2000) Insect pest management in tropical Asian irrigated rice. *Annual Review of Entomology* **45**, 549–574.

Matthews, G., Morteza Zaim, M., Yadav, R.S., Soares, A., Hii, J., Ameneshewa, B., Mnzav, A., Dash, A.P., Ejov, M., Tan, S.H., van den Berg, H. (2011) Status of legislation and regulatory control of public health pesticides in countries endemic with or at risk of major vector-borne diseases. *Environmental Health Perspectives*, **119**, 1517–1522.

Oerke, E-C. and Dehne, H-W. (2004) Safeguarding production: losses in major crops and the role of crop protection. *Crop Protection* **23**, 275–285.

Oliver, R. and Hewitt, H.G. (2014) *Fungicides in Crop Protection*. CABI Publishing, Walingford.

Parker, C. (2012) Parasitic weeds: a world challenge. *Weed Science* **60**, 269–276.

Perry, L., Adams, R.D., Bennett, A.R., Lupton, D.J., Jackson, G., Good, A.M., Thomas, SH., Vale, J.A., Thompson, J.P., Bateman, D.N. and Eddleston, M. (2014) National toxicovigilance for pesticide exposures resulting in health care contact – An example from the UK' s National Poisons Information Service. *Clinical Toxicology* **52**, 549–555.

Roberts, D.M., Buckley, N.A. Manuweera, G. and Eddleston, M. (2003) Influence of pesticide regulation on acute poisoning death in Sri Lanka. *Bulletin of the World Health Organization* **81**, 789–798.

Thomas, M.R. (2000) Pesticide usage monitoring in the United Kingdom. *The Annals of Occupational Hygiene* **45**, S87–S93.

Thundiyil, J.G., Stober, J., Besbelli, N. and Pronczuk, J. (2008) Acute pesticide poisoning: a proposed classification tool. *Bulletin of the World Health Organization* **86**, 205–209.

Tunstall J. P. and Matthews, G.A. (1966) Large scale spraying trials for the control of cotton insect pests in Central Africa. *Empire Cotton Growing Review*, **43**(3), 121–139.

Ware, G.W. (1986) *The Pesticide Book*, 5th Edition. Thomson Publications, Fresno, CA.

Way, M.J. and Heong, K.L. (1994) The role of biodiversity in the dynamics and management of insect pests of tropical irrigated rice. *Bulletin Entomological Research* **84**, 567–587.

WHO (2010) *The WHO Recommended Classification of Pesticides by Hazard and Guidelines to Classification 2009*. WHO, Geneva.

2 Approval of pesticides

Before a pesticide can be marketed, the companies developing it have to produce a large dossier of information to the regulatory authorities. Previously, each country had its own system of regulation, but the trend has been to harmonise the procedures. In Europe, registration of the active ingredient has to be completed initially with one of the Member States acting as the rapporteur for compiling the assessment of the data, prior to individual Member States registering products/formulations containing the pesticide in individual countries. Member States can only authorise the marketing and use of plant protection products after an active substance has been added to the list of approved active substances. The European Commission Regulation (EU) No. 283/2013 of 1 March 2013 sets out the data requirements for active substances, in accordance with Regulation (EC) No. 1107/2009 of the European Parliament and of the Council concerning the placing of plant protection products on the market. Active substances that do not have an unacceptable risk to people or the environment are added to the list of approved active substances, contained in the Commission's Implementing Regulation (EU) No. 540/2011. This Regulation therefore follows the European Commission's use of the precautionary principle to protect human health and the environment:

> The purpose of this Regulation is to ensure a high level of protection of both human and animal health and the environment. Particular attention should be paid to the protection of vulnerable groups of the population, including pregnant women, infants and children. The precautionary principle should be applied and ensure that industry demonstrates that substances or products produced or placed on the market do not adversely affect human health or the environment. (Anon, 2006)

"Risk" vs "Hazard"

The precautionary principle provides the legal but not scientific means of addressing uncertainty with a bias towards safety (Simon, 2014). In contrast to the dose–response data used in risk assessment to quantify the relationship between dose of a particular chemical to which an individual or population is exposed and the likelihood of adverse health effects, the precautionary principle merely identifies the possibility of a hazard when the chemical is used. As discussed later, the all-important factor is the dose to which individuals are exposed, is absorbed or ingested and interacts with a particular organ or cell.

An active ingredient may be included for a maximum period of 15 years. This can be renewed, but it can also be reviewed at any time. A product may be revoked if new evidence is obtained that in commercial use there are questions about the safety of the product. The European Food Safety Authority (EFSA) has a Pesticide Risk Assessment Peer Review group and the Panel on plant health, plant protection products and their residues (PPRs). Risk management is considered separately by the Health and Consumer DG. The approval may also be reviewed if data requirements are changed. Many older products have been withdrawn recently because of more up-to-date data requirements and the manufacturers have not considered it cost-effective to invest in further research to support continued registration. This 'regional' approach is needed in other parts of the world where individual countries do not have the trained manpower to assess all the technical data (Matthews et al., 2011). Nevertheless, some have argued that authorities in African countries need to generate data such as occupational, dietary, residential and environmental data related to the tropical conditions as they rely extensively on risk assessments from temperate regions (Utembe and Gulumian, 2014). Guidelines for registration have been published (WHO/FAO, 2010).

Each country should have legislation that provides the legal framework for pesticide use. In the USA, the Environmental Protection Agency (EPA) is responsible for pesticide regulation.

Individual European countries assess data requirements for plant protection products also in accordance with Regulation (EC) No. 1107/2009 of the European Parliament, while placing these products on the European market is according to Commission Regulation (EU) No. 284/2013 of 1 March 2013. In the UK, the Chemicals Regulation Directorate (CRD), a Directorate of the Health & Safety Executive (HSE), is responsible for ensuring the safe use of biocides, industrial chemicals, pesticides and detergents to protect the health of people and the environment.

The CRD covers the REACH (Registration, Evaluation, Authorisation and Restriction of Chemicals) Regulation, the Biocidal Products Directive (BPD) and ongoing regulatory responsibilities under the UK Control of Pesticides Regulations (COPR); Plant Protection Products Directives and Regulations; and Detergents Regulations and the EU Classification, Labelling and Packaging Regulation. The assessment of agricultural pesticides is considered in relation to three different climatic and crop growing zones in Europe to share assessments and recognise each other's authorisations. In Europe integrated pest management (IPM) is now recognised as a mandatory part of crop protection and is designed to encouraged adoption of control measures to reduce dependence on pesticides in selected major farming systems in

Europe, thus reducing the risks to human health and the environment. Greater emphasis is now on developing farming systems, to reduce reliance on single solution strategies and creating additive/synergistic interactions between IPM components. This has been accompanied by a major reduction in the number of pesticides available to European farmers. The highly toxic chemicals and most persistent pesticides have been withdrawn and older products also withdrawn if the market is insufficient to justify the cost of obtaining the data to meet current regulatory requirements. Similar actions are now being untaken by other countries, thus with FAO assistance Mozambique has taken important regulatory measures to protect its people and the environment by cancelling the registration of 79 Highly Hazardous Pesticides (HHPs). UNEP has also issued a Strategic Approach to International Chemicals Management (2014). The Strategic Approach to International Chemicals Management (SAICM) is a policy framework to promote chemical safety around the world. The objective of SAICM is to achieve the sound management of chemicals throughout their life cycle so that by 2020, chemicals are produced and used in ways that minimize significant adverse impacts on human health and the environment.

As part of the changes in registration of pesticides, it was emphasised that a range of products is needed to enable resistance management to be possible with products having different modes of action, that some products may need to be used 'off-label' on minor crops, and that any less toxic alternative to a recommended pesticide should be subject to comparative assessment and have equivalent control efficacy under variable conditions and market requirements in terms of crop safety before substitution (Anon, 2014).

In the UK, advice is provided by an independent committee, the Advisory Committee on Pesticides (ACP) (Anon, 2000), which has been a non-governmental public body (NGPB), but is now an 'expert' committee*. This committee was set up on the recommendation of a Working Party on Precautionary Measures against Toxic Chemicals, set up in 1950, with Professor Zuckerman as chairman. In 1954, the Advisory Committee on Poisonous Substances used in Agriculture and Food Storage (ACPS) covered both pesticides and veterinary medicines used on farms. Veterinary medicines were removed from the committee's role after the Medicines Act in 1968 and since 1983, membership of the committee has been independent of both the Government and industry. Under the Nolan Rules, applicants are interviewed and those that meet the criteria in terms of specialist technical experience needed on the committee are short-listed for approval by the Minister. Recommendations made by the ACP go to the relevant Ministers, who must all agree on any decisions. Advice from the ACP is also used in Northern Ireland.

*Now the Expert Committee on Pesticides (ECP).

In the UK, surveys of crops are conducted to determine pesticide usage. The main purpose of the surveys is to

- Assist policy, including assessing the economic and/or environmental implications of introduction of new active substances and the withdrawal/non-approval of pesticide products (the data reported to organisations such as the OECD and EU enabling the UK to honour international agreements); evaluating changes in growing methods and IPM where this has an impact on pesticide usage. (Pesticide usage data is required under EU Statistics Regulation (1185/2009/EC));
- Informing the pesticide risk assessment (approval) process;
- Providing information to assist targeting of monitoring programmes for residues in food and the environment.
- Responding to enquiries (e.g. Parliamentary Questions, correspondence, queries under the Freedom of Information Act).
- Informing the Wildlife Incident Investigation Scheme (WIIS) programme to help identify potential misuse of pesticides.

Each pesticide has several names. Apart from a chemical name decided by the rules of the International Union of Pure and Applied Chemistry (IUPAC), most pesticides are assigned a common name agreed by the International Organisation for Standardisation (ISO). Each company manufacturing a pesticide may have one or more formulations marketed with different trade names.

As biopesticides are now being promoted within IPM programmes, registration authorities have started to adopt a simplified system of approval. In the UK, manufacturers are advised to discuss data requirements with the CRD (Chemicals Regulation Directorate) as the 'biopesticides' are considered under four categories: (1) Pheromones and semiochemicals, (2) microbial pesticides, (3) plant extracts (assessed on a case by case basis as requirements will vary) and (4) other novel products. Registration is not required for multicellular organisms such as entomopathogenic nematodes (EPNs) and insect predators. The system has enabled the cost of registration to be lower than with conventional pesticides.

Pesticides can only be approved if they are effective, cause no serious illness through their use, nor harm anyone as a result of residues in food or drinking water that might be found following good agricultural practice (GAP). Furthermore, they should not cause any adverse effects to the environment when used according to the conditions of their approval. In order to meet these requirements, the data package is scrutinised in detail and a risk assessment made. The main problem in the EU is that implementation of the precautionary principle is not taking adequate account of the scientific data as decisions are influenced by political issues.

Table 2.1 Data requirements

Metabolism data – It is important to know what happens to the pesticide in the body, what metabolites are produced and how it is excreted.
Acute toxicity – The effect of a single dose of the active ingredient and of the product by oral, dermal and inhalation exposure.
Chronic and 'sub-acute' toxicity – The effect of exposure of the active ingredient when administered to animals over a long period. This test is usually for 2 years with rats, but a 1 year test may be needed with dogs.
Carcinogenicity – Whether the active ingredient has the potential to cause cancer when administered for a minimum of 2 years to rats, or 18 months to mice.
Genotoxicity – Assessments of the potential of the active ingredient to damage the genetic material in cells.
Teratogenicity – Whether the active ingredient can cause foetal death or malformations when administered to female animals during pregnancy.
Generation study – In relation to chronic toxicity, whether the active ingredient has a potential to impair fertility and the ability to rear young.
Irritancy – Tests are also carried out to assess whether when it is applied it will cause irritation to the skin or eyes, or has the potential to cause sensitisation, such as skin allergies. The outcome of these tests will affect the labelling of the product.

In the UK, the companies have been able to obtain an experimental approval to allow development studies on new pesticides, or new uses for existing pesticides, on a limited scale so that the scientific data on its efficacy and the residues obtained in treated crops can be determined. Unless the risk to consumers is assessed, the crop may have to be destroyed. A product may then get provisional approval, which allows commercial use while additional confirmatory data that may be needed are obtained. Provisional approval is not granted, if there are outstanding 'safety' issues.

On some occasions if a new alien pest is detected in the UK, special control measures may be needed to prevent its spread. This may require an emergency approval for a product not normally used in the UK, or is approved only on certain crops.

Pesticides are poisons, and so a key aspect of the risk analysis is the potential toxicity to humans. Much of the data required (Table 2.1) is obtained by experiments using the rat as a 'model' mammal, but a few tests may be with dogs or rabbits. The number of laboratory animals used in these tests is kept to an absolute minimum. Additional tests may be required depending on the mode of action of the pesticide and whether any effects on specific body systems, such as the nervous, immune or endocrine systems, need to studied in greater detail.

The long-term studies at a range of doses will indicate the feeding rate at which the active ingredient in milligrams per kilogram bodyweight (mg/kg bw) shows the first signs of any change in the test animals compared with untreated controls. Close observation will reveal any changes apart from ill-health that occur. Weight measurements are routine to see if there is any gain or loss. The tests will enable a dose, which defines the reference point below which no adverse symptoms occur. This is called the

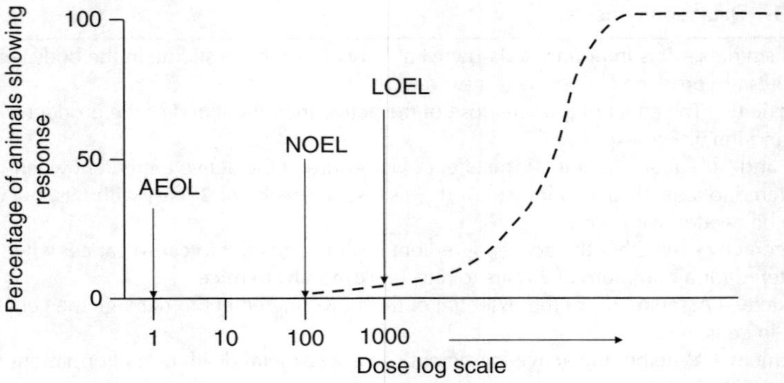

Fig. 2.1 Relative position of AEOL, NOEL and LOEL on a dosage response curve. Royal Commission on Environmental Pollution.

no-observed-adverse-effect level (NOAEL). Sometimes, reference is made to a no-observed-effect level (NOEL) (Fig. 2.1).

This reference point is then used as a starting point to derive an acceptable daily intake (ADI), which is the amount of chemical that can be consumed every day for a lifetime in the practical certainty, on the basis of known facts, that no harm will result. In calculating safe daily intakes of food by humans, a 100-fold safety or uncertainty factor has been long established to allow for differences between the animal used in tests and humans and the inter-individual variability (Renwick et al., 2000). Thus it assumes that a human may be 10× more sensitive than a test animal and a sensitive adult or child will be 10× more sensitive than the average human. These long-term tests take into account not only the active ingredient ingested but also any metabolites produced in the body, as these may in some cases be more toxic than the original pesticide. Models are used to estimate the exposure to a new active substance or rank exposure of one pesticide to others used in similar conditions. Ross et al. (2015) provide a brief overview of the range of human exposure models that are available.

Concern about the so-called cocktail effect of residues of more than one pesticide being found in a food sample has also been expressed. EFSA (2013) has begun consideration of the development of harmonised terminology and frameworks for the human risk assessment of combined exposure to multiple chemicals (chemical mixtures). As a first step, a risk assessment of combined exposure to multiple chemicals is conducted using a tiered approach for exposure assessment, hazard assessment and risk characterisation. The tiers can range from tier 0 (default values, data poor situation) to tier 3 (full probabilistic models). In the UK, a committee set up by the Food Standards Agency concluded that the risk to people's health from mixtures of residues is likely to be small. Although it is considered that children and pregnant or breastfeeding women were unlikely to be more affected by the 'cocktail effect' than most other people, it was difficult, with only limited evidence, to predict how some chemicals would interact. Further research is needed, but in the interim the default assumptions for regulatory purposes are that chemicals with different modes of action will act independently and those with the same toxic action will act additively. Where there is a possibility of an interaction, such as potentiation, adequate dose–response data is essential for interpreting dietary intake and human exposure to the mixture (Woods and Committee of Toxicity of Chemicals in Food, 2004).

On some days, people may consume much higher quantities of certain foods or they may eat some foods only at one time in the year. In view of this, another important parameter is the acute reference dose (ARfD). While similar to the ADI, it refers to the consumption of the amount of active ingredient at one meal or on one day. Short-term dietary intake was discussed by Hamilton et al. (2004), who reported the recommendations of the IUPAC Advisory Committee on Crop Protection Chemistry relating to acute dietary exposure. The value of ARfD is based on the lowest NOAEL obtained in acute toxicity or developmental toxicity tests, adjusted by an appropriate uncertainty factor. Guidance on setting the ARfD has been provided by Solecki et al. (2005).

Both ADI and ARfD relate to the ingestion of the pesticide, whereas it is the skin of those working with pesticides that is often most exposed to the pesticide. For these workers, the acceptable operator exposure level (AOEL) is the most important reference. The AOEL is set at a level of daily exposure that would not cause adverse effects in those working with the pesticide regularly over a period of days, weeks or months. It is calculated on the basis usually of short-term toxicity studies, usually up to three months' duration, but other studies may be taken into account depending on the type of chemical and pattern of usage of the pesticide. Much of the emphasis on toxicology is related to the oral route, so numerous assumptions are made in worker risk assessments. It has been suggested therefore that methods of assessing dermal absorption, including the use of human subjects, needs to be improved and that more needs to be known about interspecies pharmacokinetics to determine an appropriate toxicology study regime to reflect intermittent worker exposure (Ross et al., 2001).

Information on pesticides, including the acute oral, dermal and inhalation toxicity, the NOEL, ADI, WHO toxicity class and EC hazard index is available from *The Pesticide Manual*, a publication of the British Crop Production Council (BCPC), that is also available in electronic format. Information on the registration of pesticides is provided in the UK for agricultural use by the Pesticides Safety Directorate through the Chemicals Regulation Directorate (CRD), an agency of the Health and Safety Executive (HSE). A Guide is published that lists the active ingredients, which may be used on crops, and the pests controlled. This UK Pesticide Guide then provides a list of all the products that have been registered for each active ingredient that has approval, and for each active ingredient, it lists the uses permitted, environmental safety, health classification and safety precautions required. An electronic version allows access to updates between the annual publication of the guide. The guide also indicates extension of authorisation for minor uses (EAMUs) previously referred to as 'off-label' uses. Similar information is provided by all countries, which have a regulatory system. Further information is also available from the Internet (e.g. http://www.epa.gov/pesticides, http://www.pesticideinfo.org and http://www.pesticides.gov.uk.)

Before data reaches the ACP, it is considered by an inter-departmental secretariat (IDS) formed by members of the UK regulatory departments, who also consider data being submitted under the European legislation. The ACP also draws on the specialist knowledge of members of several other committees and panels that cover toxicology, the environment, residues and usage surveys. Minutes of ACP meetings are made available on the UK Chemicals Regulation Directorate web page.

While the developed countries have appropriate legislation and government staff to implement the registration of pesticides, this situation does not exist on the same scale in developing countries. Some countries may have only a limited staff and virtually no support, such as residues laboratories. These countries have tended to rely on whether a pesticide has been registered by the EPA or in a European country.

Some have introduced local labelling requirements such as colour coding to indicate the WHO classification. Thus in Zimbabwe, pesticides had to have a purple, red, orange or green label according to whether the product was in WHO Class I, II, III or unclassified. To buy purple or red products, which were not openly displayed by distributors, the purchaser had to ask specifically for it and was required to know about the higher toxicity of the pesticide. Guidance on harmonising labelling of chemicals is given at http://www.unece.org/trans/danger/publi/ghs/ghs_rev05/05files_e.html. Unfortunately many retailers in tropical countries have not received adequate training and supply highly toxic pesticides, often repackaged with inadequate labelling (Lekei et al., 2014).

Responsible registration of pesticides has limited the use of the most toxic pesticides in many countries or restricted their use to fully trained operators. In contrast, a lack of enforced regulation in many other areas of the world has allowed untrained people access to highly toxic pesticides. Doctors in developing countries have had to deal with many cases of poisoning as a result of highly toxic insecticides being applied without adequate protective clothing (Ngowi et al., 2001). Unfortunately, many deaths caused by pesticides are due to suicides, where people have over-used them and then had insufficient income from their crop to repay debts. This has been particularly noted in parts of Asia.

One of the main causes of pesticide poisoning among farmers in the tropics is due to exposure to the concentrated formulated product when preparing sprays. Measuring out small quantities of pesticide to apply with manually operated equipment exposes the operators' hands unless suitable packaging such as sachets are used, or containers have a built-in measure (Fig. 2.2) or farmers are provided with ancillary equipment to allow measurement of small quantities safely. Labels on pesticide containers should indicate what personal protection equipment (PPE) is required. Labels should also have pictograms to indicate PPE and other information (Fig. 2.2e). However, it has been argued that standardised approaches such as the pictograms are not universally understood, so users of pesticides need to be trained in a language they understand to encourage compliance with

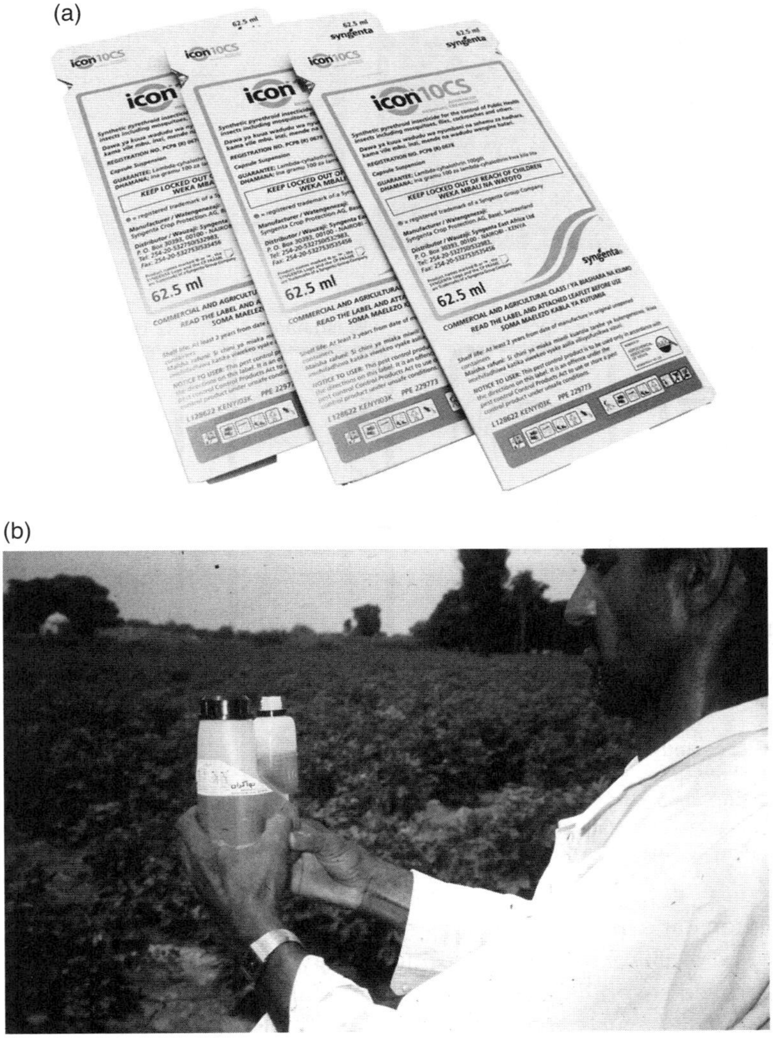

Fig. 2.2 Different containers. (a) Sachets (Photograph courtesy of Syngenta). (b) Container with built-in measure (Photograph by Graham Matthews). (*See insert for color representation of the figure.*)

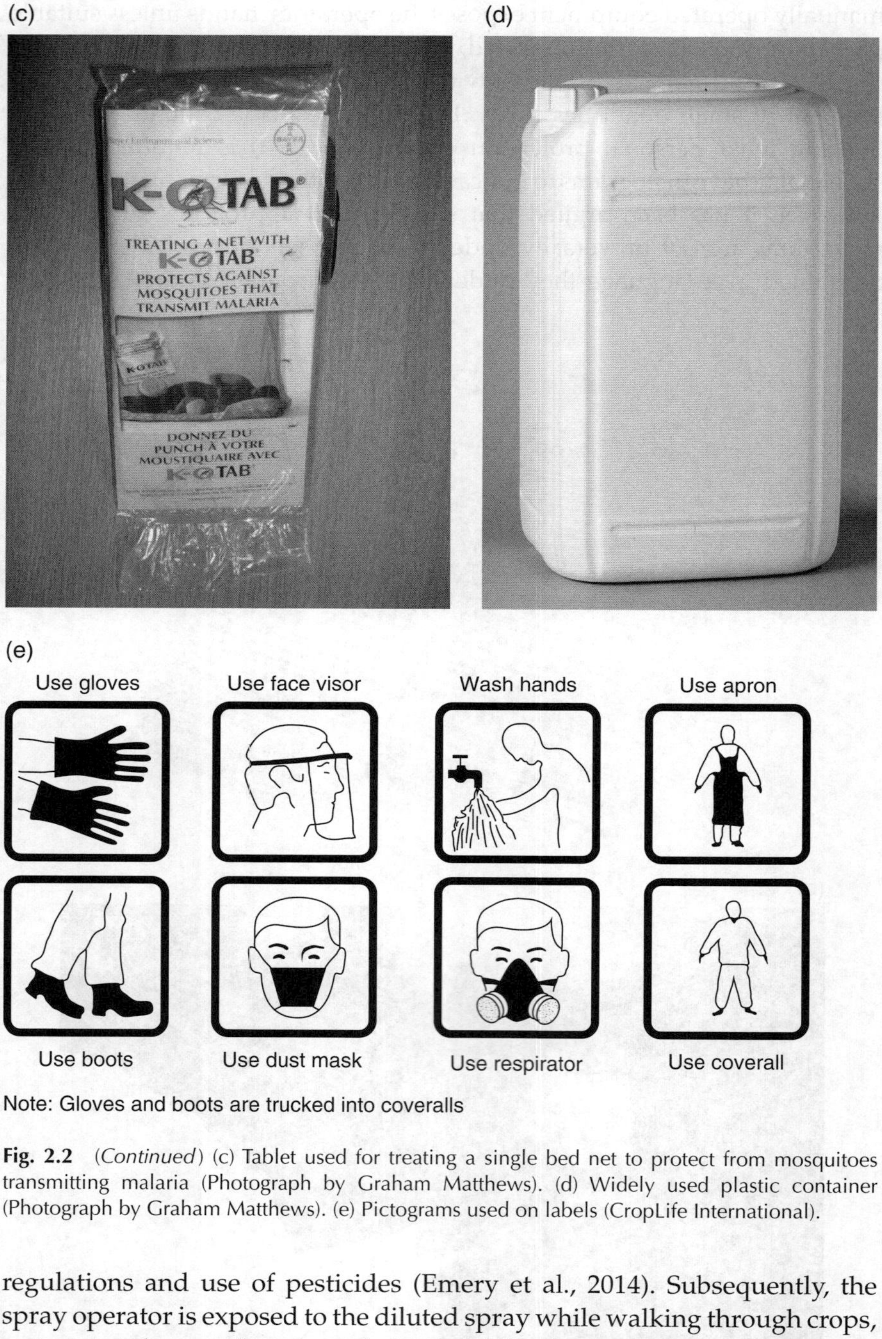

Fig. 2.2 (*Continued*) (c) Tablet used for treating a single bed net to protect from mosquitoes transmitting malaria (Photograph by Graham Matthews). (d) Widely used plastic container (Photograph by Graham Matthews). (e) Pictograms used on labels (CropLife International).

regulations and use of pesticides (Emery et al., 2014). Subsequently, the spray operator is exposed to the diluted spray while walking through crops, so improved coveralls are needed suitable for hot climates to provide protection of areas, such as the lower legs, which are most exposed to a spray. Spray operator training and certification will need to be increased, but this is a major task in tropical countries with very many small farms.

Clearly, with an increasing world population to be fed, pesticides will continue to be an important tool in IPM/ICM programmes, but for this to be more acceptable greater efforts are needed to minimise ill health due to operator exposure as well as minimising environmental pollution.

Retrospective assessment

Perhaps, the most famous correlation of ill health and human activity was the link between lung cancer and smoking tobacco. An epidemiological study revealed that people exposed to the contents of tobacco smoke including nicotine were undoubtedly more likely to suffer from lung cancer. Such epidemiological studies are more complex with pesticides as the route and period of exposure to different pesticides is quite varied compared with direct inhalation of tobacco smoke by individuals. It has been suggested that people are exposed to pesticides through ingesting residues in food (see Chapter 7), through inhaling air contaminated by spray or by direct dermal exposure when using pesticides. However, apart from those working in pesticide manufacturing facilities and users of pesticides, the quantities in each case are extremely low and mixed with many other chemicals. Air, especially in towns and cities is also contaminated with vehicle exhausts, while many foods naturally contain many different chemicals.

Nevertheless, epidemiologists do study disease patterns to establish whether there are causal factors. One type of study is a cohort design, in which a group of people share a common characteristic. A study might include a group of certified spray applicators. At the start of the study, participants should initially be free of the disease under investigation, but their subsequent exposure to pesticides and health patterns are followed. The frequency of disease incidence between exposed and unexposed populations is then analysed to assess whether the exposure was the cause of ill health. If the group had been exposed at some time in the past, then a retrospective cohort study is carried out, based on the records of individuals and other relevant data, such as air monitoring data. Cohort studies require participation of a large number of people over a long period and are therefore expensive to conduct. In the UK, a Pesticide Users' Health Study (PUHS) was set up and participants have been followed since 1987. Frost et al. (2011) reported that users were generally healthier than the national population, but may have more non-melanoma skin cancer, testicular cancer and multiple myeloma. However, mortality due to injuries caused by machinery was significantly more than expected for men. In the USA a study found significant associations between cutaneous melanoma and certain pesticides, including maneb/mancozeb and parathion (Dennis et al., 2010).

Where a study investigates, for example, the occurrence of a birth defect in a group of children, a case-control design can be used. Exposure data is sought from existing records or detailed questionnaires completed by the subjects, or next-of-kin, to compare the frequency of exposure with a similar unexposed control group, adjusted to allow for other factors that might have influenced the disease. Calculation of the ratio of disease incidence among those who were exposed, or non-exposed with a similar group without the defect, who were exposed or not, can give an indication of whether rate of defect incidence was higher or not. Recall of details of exposure may not always be reliable if those suffering from a defect or disease are motivated to participate in the study.

Research by medical doctors and epidemiologists outside the laboratory is important as it can provide information that cannot be predicted from tests on non-human animals. Information from multiple exposures under real-world conditions to a much larger population is crucial in confirming the verdicts from regulatory authorities. Nevertheless, care is needed in interpreting epidemiological studies unless the study has sound exposure data. Unfortunately, when news media report such studies, some aspects are over emphasised without any scientific disclaimers given by the original authors, leading to sensational comments and fear mongering.

In Canada, an attempt was made to draw together information from selected published reports of epidemiological studies related to a number of human diseases and alleged pesticide exposure. The report (Sanborn et al., 2004) acknowledged that epidemiology studies are difficult to interpret because of biases and confounding factors, making it difficult to establish any link between pesticide exposures and illnesses. This is especially as people encounter other chemical and physical environment effects that may have been responsible for the illness. A weakness of many studies is the use of surrogate information (sales data, crops grown, recall of what was applied) in the absence of being able to quantify what levels of pesticides the individuals were actually exposed to, and when these exposures occurred. Thus in the absence of actual exposure data, it is not possible to assess whether pesticides could be the cause, but the general observation that exposure to pesticides should be curtailed as much as possible, especially for children, is undoubtedly correct. However, it is extremely difficult to unravel the causes of possible chronic effects, particularly as recall of exposure is very difficult and often vague.

In the UK, the Medical and Toxicological Panel of the Advisory Committee on Pesticides scrutinises annually the published papers on pesticides and human health to assess whether any regulatory action is required.

Environmental aspects

Potentially there is risk of pesticides adversely affecting all non-target organisms. Much depends on the toxicity of the pesticide, the application rate and how it persists in the environment. Initially the fate and behaviour of the pesticide is assessed with calculations of the predicted environmental concentration (PEC). In the USA, the PEC is referred to as an estimated environmental concentration (EEC). These environmental concentrations are calculated for soil, water, sediment and air. As it would be extremely expensive to measure the concentration of pesticide in many different situations, models are used to predict the PEC based the physical properties of the chemical and validated in some situations by actual measurements. An example of PEC_{water} (Table 2.2) shows the decrease with distance and over time.

In the approval system, data on the effect of key non-target species is required to compare it with the PEC. The toxicity exposure ratio (TER) is used to determine whether the risk to the organism is acceptable or not. The TER is calculated from the LC_{50} or equivalent measure of the susceptibility of an organism divided by the PEC relevant to the situation in which the organism is living. Thus the PEC_{water} is used to assess the TER for fish. A TER for <100 for acute risk to fish indicates a need for detailed higher tier risk assessment. For chronic risk, the TER is <10. In the USA the calculated risk quotient is the inverse of TER, that is the PEC is divided by the indicated toxic dose. The assessment of acute toxicity to different non-target organisms is confined to selected surrogate organisms (Table 2.3). These tests are

Table 2.2 Predicted environmental concentration (PEC) values (μg/l) in water for a pesticide

Distance to water (m)	Days after treatment		
	0	7	14
0	530	450	390
5	50	45	40
10	25	20	15
50	1	0.9	0.8

Table 2.3 Three levels of tests on non-target organisms

	Tier 1	Tier 2	Tier 3
Avian species	Acute toxicity	Reproduction test	Field test
E.g. Bobwhite quail	LD_{50}		
Freshwater fish	Fish LC_{50}	Effects on spawning	Fish life cycle study
E.g. Rainbow trout or minnows			
Aquatic invertebrate	Invertebrate EC_{50}	Full life cycle	Simulated field test
E.g. Daphnia, shrimps			
Non-target invertebrates	Acute LD_{50}	Effect of residues on foliage	Field test for pollination
E.g. Honey bee and earthworms			
Terrestrial plants	Seed germination		
E.g. various crops			
Aquatic plants			
E.g. Algae	Plant vigour		

Table 2.4 Comparison of TER for two insecticides to fish when using two application rates

	Recommended dose		Reduced dose	
Pesticide	A	B	A	B
PEC_{water} (mg/l)	0.08	0.08	0.045	0.045
Fish acute toxicity LC_{50} (mg/l)	8.5	0.05	8.5	0.05
TER	106.25	0.625	188.9	1.1
Fish chronic NOEC (mg/l)	0.4	0.015	0.4	0.015
TER	5.0	0.17	8.88	0.33

followed by more specific tests relevant to the life cycle of the organism and the way in which the pesticide will be applied. For example, aldicarb is an extremely toxic insecticide and nematicide that was used effectively in many countries, but it has been phased out in the EU. This insecticide was only used as a low-percentage granule for soil treatment, but there were still instances of bird mortality where granules remained on the soil surface. The final tier is a simulated or real field test.

If two different pesticides are compared (Table 2.4) using Tier 1 data and the risk is not acceptable as shown by Pesticide B, then further data are needed from Tier 2 or 3 to see if there is any way the pesticide can be used to reduce its toxicity to fish. This may be by changes in formulation, reduced dosage, if still effective against the pests, or a wider buffer zone to protect water. Recognising that Pesticide B is acutely toxic to fish (Table 2.4), the TER for acute toxicity at a reduced application rate and a lower PEC_{water} is smaller; but in this example, it is still not acceptable so further evaluation would be needed at Tier 2, especially as the chronic toxicity TER is unacceptable.

Similar data are generated for all the non-target organisms evaluated. The PEC_{soil} can vary with different crops, depending on the application rate and frequency of application and persistence of deposits in the soil. As the sensitivity of soil inhabitants can also vary, for example between species of earthworms, the TER needs to be calculated to relate to the particular circumstances where the pesticide will be used.

Changes in evaluating pesticides in relation to non-target arthropods have been proposed, that is the European Standard Characteristics of Non-Target Arthropod Regulatory Testing (ESCORT), which considers both in-field and off-field effects for spray treatments. This is likely to be most important for certain pesticides, such as insect growth regulators. Information from a workshop has been published (ESCORT 3, Alix et al., 2012) in a book and focuses on four key topics – level of protection and testing scheme, off-crop environment, recovery and field studies. It provides conclusions and recommendations that also include previous ESCORT programmes.

With concern about toxicity of insecticides to bees and other pollinators, Fischer and Moriarty (2014) report the regulatory risk assessment and decision-making processes needed to determine whether a pesticide can be approved.

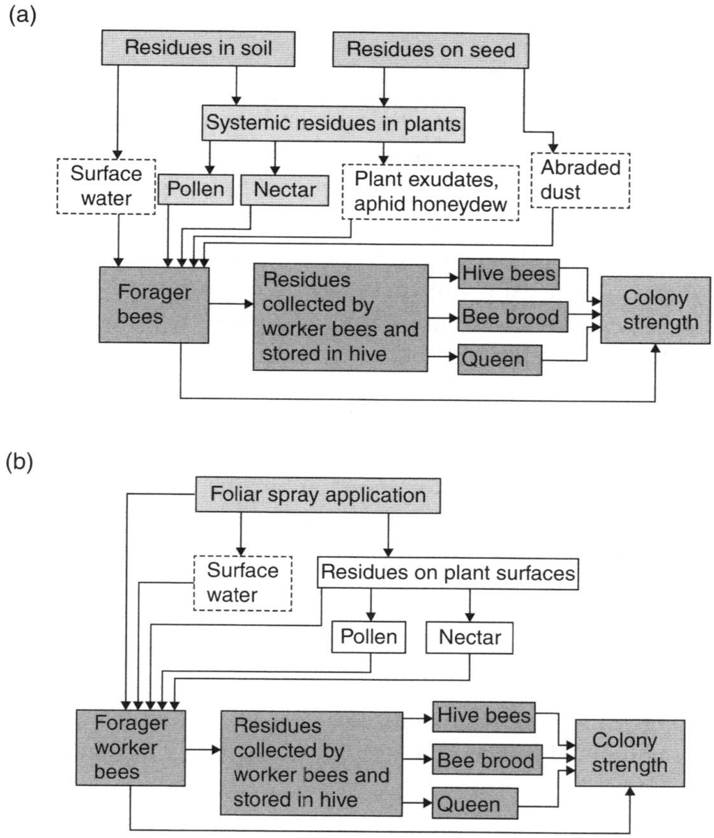

Fig. 2.3 Diagram to show potential routes of exposure of bees to insecticides applied (a) as soil or seed treatment and (b) as a foliar spray (Fischer and Moriarty, 2014). (Reproduced with permission of John Wiley & Sons).

They provide a detailed description of a proposed overall risk assessment scheme to cover both spray, soil and seed treatments (Fig. 2.3), in view of the difficulties in determining the extent to which pollinators are exposed to pesticide deposits. The risk assessment also needs to consider whether commercial use of bees as pollinators is spreading pathogens affecting the health not only of honey bees but also wild pollinators (Manley et al., 2015).

Companies developing new molecules assess their future prospects by carrying out comprehensive risk assessment to ensure that their investment will lead to a commercially registered product. One example of a risk assessment for a novel insecticide spinosad was reported by Cleveland et al. (2001). The trend has been towards new low-dosage insecticides to control sucking pests to complement the commercial expansion of major GM crops effective against lepidopteran pests. Another trend is the move by agrochemical companies to include bio-pesticides in their range of products.

Endocrine disrupters

Some organisations concerned about safety have called for banning of certain pesticides, which are referred to as endocrine disrupters, or environmental estrogens. The argument is that such chemicals could adversely affect hormone balance, or disrupt their action regulating the normal function of organs (Colborn et al., 1993). In the media, claims of an epidemiological study that sperm counts of humans had declined by almost 50% over the past 50 years due to exposure to synthetic chemicals hit the headlines. Immediately pesticides were implicated, despite our exposure to a whole range of different chemicals in the environment.

The EPA defines them as chemicals, both from natural and man-made sources, which interfere with the synthesis, secretion, transport, binding action or elimination of natural hormones in the body. WHO defines an endocrine disruptor as an exogenous substance or mixture that alters function(s) of the endocrine system and consequently causes adverse health effects in an intact organism, or its progeny, or (sub) populations. Some drugs bind to hormone receptors and elicit a pharmacological response, for example 'Tamoxifen' prescribed to treat certain types of breast cancer. As with other chemicals any effect depends when and how large is the dose present in the body. Below a threshold dose, there will be no effect and at very small doses above the threshold there may even be a beneficial effect. It is only at large doses that adverse effects would occur.

Many endocrine disruptors are thought to mimic hormones as their chemical properties are similar to hormones, and this allows them to bind to hormone-specific receptors on the cells of target organs. Like other chemical groups, endocrine disruptor chemistry and potency varies. The generally low potency of most endocrine disruptors means that a higher dose is required to obtain the same response, as the hormone that they mimic.

A large number of chemicals for ED activity are being studied in the USA by their Endocrine Disrupter Screening programme (EDSP), while in Europe, the Centre for Ecotoxicology and Toxicology of Chemicals proposed a tiered approach for the ecological risk assessment of endocrine disruptors, integrating exposure and hazard. WHO and the United Nations Environment Programme has called for more research to fully understand the associations between EDCs and the risks to health of human and animal life. Data from multi-generation animal studies would provide strong evidence if a pesticide had the potential to act as an endocrine disruptor, especially as any effects on reproduction are assessed. Where specific studies have been made, for example with alligators and birds as well as rats, the effects on reproduction have been achieved only at very high doses. This has led to speculation by some toxicologists that exposure to lower doses of some chemicals in the environment could be unacceptable not take into account the normal excretion and breakdov Nohynek et al. (2013) reported that overall despite 20 year human health risk from exposure to low concentration: chemical substances with weak hormone-like activities remains an unproven and unlikely hypothesis.

Of the many of the pesticides with a wide range of properties, toxicity and persistence that have been claimed to have reproductive and endocrine disrupting effects, perhaps the most discussed in the media has been the herbicide atrazine. Due to uncertainties associated with laboratory and field studies prior to 2003, additional studies were requested to determine if exposure to atrazine affected amphibian gonadal development, using *Xenopus laevis* as the test species for a Tier 1 study. A review in 2007 by the EPA, based on nineteen studies showed no effects of atrazine on amphibian gonadal development. However, the EPA is criticised arguing that its risk assessment is compromised as there is a conflict of interest when most of the data used in an assessment are produced by a pesticide's manufacturer (Boone et al., 2014).

Many of the pesticides considered to be endocrine disruptors are no longer approved, but within Europe, Regulation (EC) 1107/2009 on plant protection products now says that active substances, safeners and synergists shall only be approved, if they are not considered to have "endocrine disrupting properties that may cause adverse effects" in humans or non-target organisms. The European Commission, DG Environment therefore commissioned a report on the State of the Art Assessment of Endocrine Disrupters (Kortenkamp et al., 2011). Among the recommendations were that the legislation needed to be updated with enhanced validated and internationally recognised test methods to assess whether endocrine disruption was due to pesticides, together with guidance documents to interpret test data.

Approval in relation to efficacy

Manufacturers of new pesticides have to demonstrate that when the product is used as recommended it is effective against the pests for which it will be marketed. The aim is to stop application of an ineffective pesticide and avoid unnecessary addition of another chemical into the environment. Data are required to support the label claims, especially the recommended dose, show that the pesticide does not cause any damage to the crop or adversely affect yields and that it is not so persistent that it could cause damage to a subsequent crop. New pesticides require a high level of activity against the pests in a series of trials on crops on which the pesticide is intended to be used. It has been argued that within an IPM programme, there may be a eed for certain products which may be less active against a pest, but nevertheless contribute to the overall control strategy, for example by not adversely affecting biological control agents.

As part of this data package, consideration is now required on a resistance management policy, should the pest become resistant to the pesticide or a chemical with a similar mode of action. Often this requires the label to state the maximum number of applications or maximum amount of active ingredient applied per season in conjunction with similar or complementary products. Industry has established specialist groups to assess the occurrence of resistance in insects, fungi, weeds and rodents, and develop strategies to offset this problem, so that appropriate information is available both within the agrochemical industry and to growers. A key concern of industry is that growers do not use less than the label recommended rate. The label rate is established from the results of many trials and represents a robust rate that will be effective under a wide range of conditions. The label rate with GLP is set to satisfy retail outlets that produce should not exceed a maximum residue level (MRL). In some countries the user is bound to apply the label rate, but in some circumstances it will be too high and if a farmer applies a pesticide accurately at the right time, a lower-than-label dose will be effective. There is no evidence that using a low-dose rate will increase the incidence of resistance to a pesticide, whereas using too high a dose is more likely to select a resistant pest population. Furthermore, reduced rates of pesticides fit IPM better provided natural enemy survival is sufficient to regulate survivors of a chemical treatment.

Operator proficiency

In the UK, it is mandatory for spray operators to have passed a practical test arranged by the National Proficiency Test Council (NPTC), now City & Guilds Land Based Services. The National Register of Sprayer Operators (NRoSO) has over 20,000 members and ensures ongoing training and continued professional development (CPD). The training provides instruction on the label and an understanding how to calibrate their equipment. It was thought that with fully trained operators, they would ensure their equipment was fully operational, but as with motor vehicles, the operator test is not sufficient. Following a mandatory equipment test introduced in other European countries, there is a National Sprayer Testing Scheme (NSTS), under which 15,427 sprayers which are responsible for treating 94.8% of the sprayed area were tested in 2013–2014, in the UK. Testing revealed that repairs were required on 60.7% of machines, due to leaks and drips, worn hoses and nozzles and faulty pressure gauges, emphasising the value and importance of annual testing. Sprayer testing is now an accepted part of UK agriculture and is a requirement of the major crop assurance schemes as well as supermarket protocols. The results of a survey of farm sprayer practices in the UK were summarised by Garthwaite (2004). See also Annual Report of the Pesticide Forum. Similar schemes for inspection of sprayers operate in other European countries (Braekman and Sonck, 2004) and in many other parts of the world.

Storage and waste management

The agrochemical industry has now promoted and implemented a life-cycle, or stewardship, approach to the management of crop protection products, including good storage and stock control to avoid having obsolete sticks. Guidance on storing pesticides in the UK is provided by the Health and Safety Executive (http://www.hse.gov.uk/pubns/ais16.pdf). A key part of this policy is to collect, handle and safely dispose of any waste resulting from their use. In particular, the disposal of used pesticide containers when they are empty is also required by legislation in Europe and other parts of the world. Previously small plastic containers, after they had been triple rinsed, could be incinerated on the farm. Now it is recognised that a sustainable container management programme is needed which takes into consideration local regulations, collection system, method of disposal, the recycling of materials and financial arrangements (FAO, 2008; Denny, 2013; Jones, 2014). Since 2005, the quantity of plastic containers that have been collected has more than doubled (Fig. 2.4). In Brazil, the Campo Limpo System involving distributors, cooperatives and government authorities was set up in 2002 to ensure the environmentally correct disposal of used containers from approximately five million farmers (Rando, 2013).

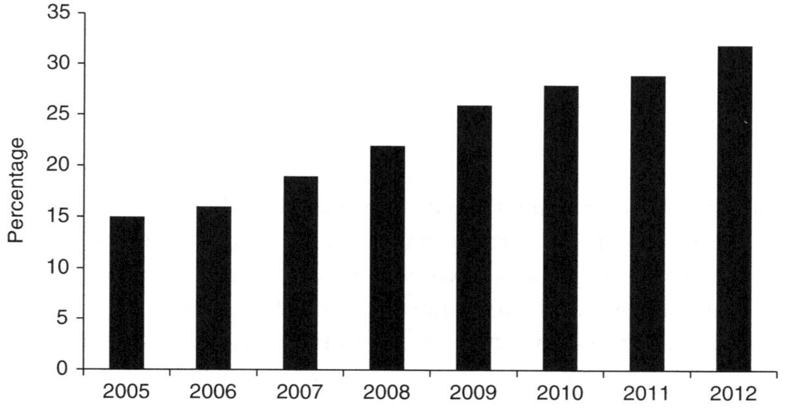

Fig. 2.4 Estimated percentage of empty containers collected globally (Adapted from Jones, 2014).

(a)

(b)

Fig. 2.5 (a) FasTrans closed transfer system on container. (b) FasTrans system showing connection to sprayer (Wisdom Systems – eziserv Ltd.). (*See insert for color representation of the figure.*)

Manufacturers of plant protection products (PPPs) have increasingly embraced the use of multi-trip containers that include a specialised connection valve. These containers have been readily adopted in some countries and market sectors, and this connection interface between container and application equipment has become an industry standard for PPPs. These containers form the basis of a closed transfer system. One system known as the 'FasTran' (Fig. 2.5) is a venturi-based extraction and metering system added to a standard sprayer and is used on about 8000 farms in Germany where BASF provide their top 25 products in compatible returnable and refillable containers. The system is also used in tropical countries on large estates, such as banana plantations. This is due to the industry recognising that the key risk for the operator is handling the concentrate at the mixing and loading site. An additional benefit that this type of packaging delivers is that it removes the need to clean and dispose of the packaging, and this increases operational and resource efficiency too. This approach has proven the benefit of closed systems however returnable containers are limited to particular markets due to the associated cost of the container and the reverse logistics of returning and managing the containers. With increasing attention now being directed towards 'negligible exposure' new methods of increasing operator safety and environmental protection are being introduced to deliver closed transfer performance for the majority of packaging and application systems in the market. Closed transfer systems greatly reduce spillages and leakages of concentrate at the mixing site where product can often enter surface drains and water systems. The use of closed transfer systems for granule application is referred to Chapter 3.

In some places, manufacturers are supporting the use of multi-trip containers. The use of the latter with closed transfer systems is referred to also in Chapter 3. Problems of container are greater in remote tropical countries with smaller farms, where containers have a value for other purposes. Two hundred litre drums are still the most economical method of long-range transport, but local repackaging is needed to ensure small scale farmers have their pesticides in appropriate quantities for use on small areas. Water soluble sachets, wrapped to prevent water access until used and containing sufficient for one knapsack sprayer are ideal, but tend to be expensive. Sachets that are not water soluble are now used in some areas and avoid the wastage that can occur once larger packaging has been opened. Safe disposal of the empty sachets still needs to be done carefully.

Apart from disposal of used containers, a need to remove stocks of pesticides that could no longer be used – the obsolete stocks – became apparent when organochlorine insecticides, such as dieldrin were no longer permitted to be used for locust control. FAO established a programme on prevention and disposal of obsolete stocks in 1994, strengthened by adoption of the Stockholm Convention in 2001. The Africa Stockpiles programme made slow progress and in 2006 FAO launched its Pesticide Management system which provides a web-based database listing inventories to assist in disposal or re-use of stocks. CropLife International has a programme of safeguarding stocks and with other partners is involved in the export and incineration of obsolete stocks (Dollimore and Schimpf, 2013).

This chapter has indicated that approval of pesticides is now regulated throughout Europe and only a limited number of active ingredients are then listed for use. Many pesticides are not approved as they are considered too toxic, or they are too persistent and liable to build up in food chains, or are unacceptable in the environment. Individual Member States then approve products that contain a listed active ingredient, following detailed examination of a vast amount of information to ensure that its use does not incur an unacceptable risk to humans and the environment. In the USA, the Environment Protection Agency oversees pesticide registration and similar programmes exist in other countries, but many do not have the resources to regulate their use sufficiently so there is a trend towards regional registration.

References

Alix, A., Bakker, F., Barrett, K., Brühl, C.A., Coulson, M., Hoy, S. Jansen, J.P., Jepson, P., Lewis, G., Neumann, P., Süßenbach, D. and van Vliet, P. (Eds.) (2012) *ESCORT 3 Linking Non-Target Arthropod Testing and Risk Assessment with Protection Goals*. Society of Environmental Toxicology and Chemistry, Pensacola, FL.

Anon (2000) A Guide to Pesticide Regulation in the UK and the Role of the Advisory Committee on Pesticides. DEFRA. Available at: www.pesticides.gov.uk/committees/acp/acpgui1.pdf (accessed on 6 May 2015).

Anon (2006) Proposal for a Regulation of the European Parliament and of the Council Concerning the Placing of Plant Protection Products on the Market. COM (2006) 388 final. Brussels.

Anon (2014) Comparitive Assessment and Substitution. Guide for UK Applicants for Plant Protection Product Authorisation. Available at: http://www.pesticides.gov.uk/Resources/CRD/Migrated-Resources/Documents/U/UK-Comp-Assess-Guidancev2.pdf (accessed on 6 May 2015).

Boone, M.D., Bishop, C.A., Boswell, L.A., Brodman, R.D., Burger, J., Davidson, C., Gochfield, M., Hoverman, J.T., Neuman-Lee, L.A., Relyea, R.A., Rohr, J.R., Salice, C., Semlitsch, R.D., Sparling, D. and Weir, S. (2014) Pesticide regulation amid the influence of industry. *BioScience* **64**, 917–922.

Braekman, P. and Sonck, B. (2004) Accreditation according to EN 4504 as a guarantee for a correct, reliable and objective mandatory inspection of sprayers in Flanders, Belgium. *Aspects of Applied Biology* **71**, 35–40.

Cleveland, C.B., Mayes, M.A. and Cryer, S.A. (2001) An ecological risk assessment for spinisad use on cotton. *Pest Management Science*, **58**, 70–84.

Colborn, T, vom Saal, F.S. and Soto, A.M. (1993) Developmental effects of endocrine-disrupting chemicals in wildlife and humans. *Environmental Health Perspectives* **101**, 378–384.

Dennis, L.K., Lynch, C.F., Sandler, D.P. and Alavanja, M.C.R. (2010) Pesticide use and cutaneous melanoma in pesticide applicators in the Agricultural Health Study. *Environmental Health Perspectives* **118**, 812–817.

Denny, R.L. (2013) Designing and implementing effective pesticide container stewardship programmes. *Outlooks on Pest Management*, **24**, 244–247.

Dollimore, L. and Schimpf, W. (2013) Obsolete pesticide stocks – the past 25 years, lessons learned and observations for the future. *Outlooks on Pest Management*, **24**, 251–256.

Emery, S.B., Hart A., Butler-Ellis, C., Gerritsen-Ebben, M.G., Machera, K. Spanoghef, P. and Frewer, L.J. (2014) A review of the use of pictograms for communicating pesticide hazards and safety instructions: implications for EU policy. *Human and Ecological Risk Assessment* **21**, 1062–1080.

EPA (2007) *White Paper on the Potential for Atrazine to Affect Amphibian Gonadal Development*. Office of Pesticide Programs, Environmental Fate and Effects Division, Washington, DC.

European Food Safety Authority (EFSA) (2013) International frameworks dealing with human risk assessment of combined exposure to multiple chemicals. *EFSA Journal* **11**, 3313.

FAO (2008) Guidelines on Management Options for Empty Pesticide Containers. FAO, Rome.

Fischer, D. and Moriarty, T. (2014) *Pesticide Risk Assessment for Pollinators*. Wiley Blackwell, Chichester.

Frost, G., Brown, T. and Harding, A-H. (2011) Mortality and cancer incidence among British agricultural pesticide users. *Occupational Medicine* **61**, 303–310.

Garthwaite, D.G. (2004) Summary of the results of a survey of current farm sprayer practices in the United Kingdom. *Aspects of Applied Biology* **71**, 19–26.

Hamilton, D., Ambrus, A., Dieterie, R., Felsot, A., Harris, C., Petersen, B., Racke, K., Wong, S-S., Gonzalez, R., Tanaka, K., Earl, M., Roberts, G. and Bhula, R. (2004) Pesticide residues in food – acute dietary exposure. *Pest Management Science* **60**, 311–339.

Jones, K.A. (2014) The recycling of empty pesticide containers: an industry example of responsible waste management. *Outlooks on Pest Management* **25**, 183–186.

Kortenkamp, A., Martin, O., Faust, M., Evans, R., McKinlay, R., Orton, F. and Rosivatz, E. (2011) State of the Art Assessments of Endocrine Disrupters. Final report. Available at: http://ec.europa.eu/environment/chemicals/endocrine/pdf/sota_edc_final_report.pdf (accessed on 29 March 2015).

Lekei, E.E., Ngowi, A.V. and London, L. (2014) Pesticide retailers' knowledge and handling practices in selected towns of Tanzania. *Environmental Health* **13**, doi:10.1186/1471-2458-14-389.

Manley, R., Boots, M. and Wilfert, L. (2015) Emerging viral disease risk to pollinating insects: ecological, evolutionary and anthropogenic factors. Journal of Applied Ecology, doi:10.1111/1365-2664.12385.

Matthews, G., Zaim, M., Yadav, R.S., Soares, A., Hii, J., Ameneshewa, B., Mnzava, A., Dash, A.P., Ejov, M., Tan, S.H. and Van den Berg, H. (2011) Status of legislation and regulatory control of public health pesticides in countries endemic with or at risk of major vector-borne diseases. *Environmental Health Perspectives* **119**, 1517–1522.

Ngowi, A.V.F., Maeda, D.N., Wesseling, C., Partenen, T.J. Sanga, M.P. and Mbise, G. (2001) Pesticide-handling practices in agriculture in Tanzania: observational data from 27 coffee and cotton farms. *International Journal of Occupational and Environmental Health* **7**, 326–332.

Nohynek, G.J., Borgert, C.J., Dietrich, D. and Rozman, K.K. (2013) Endocrine disruption: fact or urban legend? *Toxicology Letters* **223**, 295–305.

Renwick, A.G., Dome, J.L. and Walton, K. (2000) An analysis of the need for an additional uncertainty factor for infants and children. *Regulatory Toxicology and Pharmacology* **31**, 286–296.

Rondo, J.C.M. (2013) The campo limpo system reverse logistics for empty containers of crop protection products. *Outlooks on Pest Management* **24**, 273–275.

Ross, J.H., Driver, J.H., Cochran, R.C., Thongsinthusak, T. and Krieger, R.I. (2001) Could pesticide toxicology studies be more relevant to occupational risk assessment? *The Annals of Occupational Hygiene* **45**(Suppl. 1), **2001**, S5–S17.

Ross, J., Driver, J., Lunchick, C. and O'Mahony, C. (2015) Models for estimating human exposure to pesticides. *Outlooks on Pest Management* **26**, 33–37.

Sanborn, M., Cole, D., Kerr, K., Vakil, C., Sanin, L.H. and Bassil, K. (2004) *Pesticides Literature Review*. Ontario College of Family Physicians, Toronto.

Simon, T. (2014) *Environmental Risk Assessment – A Toxicological Approach*. CRC Press, Baton Roca, FL.

Soleki, R., Davies, L., Dellarco, V., Dewhurst, I., van Raaij, M. and Tritscher, A. (2005) Guidance on setting of acute reference dose (ARfD) for pesticides. *Food and Chemical Toxicology* **43**, 1569–1593.

UNEP (2014) Information Note on Highly Hazardous Pesticides. SAICM/OEWG.2/10. Food and Agriculture Organization of the United Nations, Rome.

Utembe, W. and Gulumian, M. (2014) Challenges and research needs for risk assessment of pesticides for registration in Africa. *Human and Ecological Risk Assessment* **21**, 1518–1541.

WHO/FAO (2010) *Guidelines for the Registration of Pesticides, International Code of Conduct on the Distribution and Use of Pesticides*. World Health Organisation and Food and Agriculture Organisation, Geneva/Rome. Available at: http://whqlibdoc.who.int/hq/2010/WHO_HTM_NTD_WHOPES_2010.7eng.pdf (accessed on 29 March 2015).

Woods, H.F. and Committee of Toxicity of Chemicals in Food (2004) *Risk Assessment of Mixtures of Pesticides and Similar Substances*. Consumer Products and the Environment Department of Health, London.

3 Application of pesticides

Pesticides are used in a wide range of environments. In most cases, the active ingredient is formulated so it can be diluted with water and applied by forcing the liquid through a very small opening in a nozzle to form a spray that is targeted at the intended surfaces. The process of delivery of the spray from a nozzle to the target is very complex (Figure 3.1). Equipment to do this originated in France in the late nineteenth century, when farmers started to spray Bordeaux mixture (a copper fungicide) on their vines (Lodeman, 1896). Initially a hand-operated pump was used as part of a small tank carried on the user's back; the first knapsack sprayer. Soon larger versions that were horse drawn were designed and were the forerunners of the tractor-operated equipment used by farmers today.

Apart from pesticide use on farms, these products are also used in homes through pressure packs (often called aerosol cans) and in gardens with compression or knapsack sprayers. In the tropics, the inside walls of houses may be sprayed with insecticide to control mosquitoes and other disease vectors, although bed nets fabricated with fibres impregnated with insecticide are also used to reduce transmission of malaria. Vector control also involves treatment of the mosquito-breeding sites, such as water pools, ditches, and so on, with an insecticide aimed at killing the larvae, and by space treatments with vehicle mounted or aerial equipment.

This chapter briefly describes the main types of pesticide application equipment in order to place the information in the following chapters into context.

Hydraulic sprayers

Worldwide most pesticides are applied through hydraulic sprayers, albeit they vary in size and complexity (Figs. 3.2–3.5). From the smallest knapsack to aerial equipment, the main parts are a tank, a pump and a set of nozzles, interconnected by pipes/hoses and control valves. More detailed information is given in other books on different types of sprayers (Matthews et al., 2014) and their use on different crops that are treated with pesticides

Pesticides: Health, Safety and the Environment, Second Edition. G.A. Matthews.

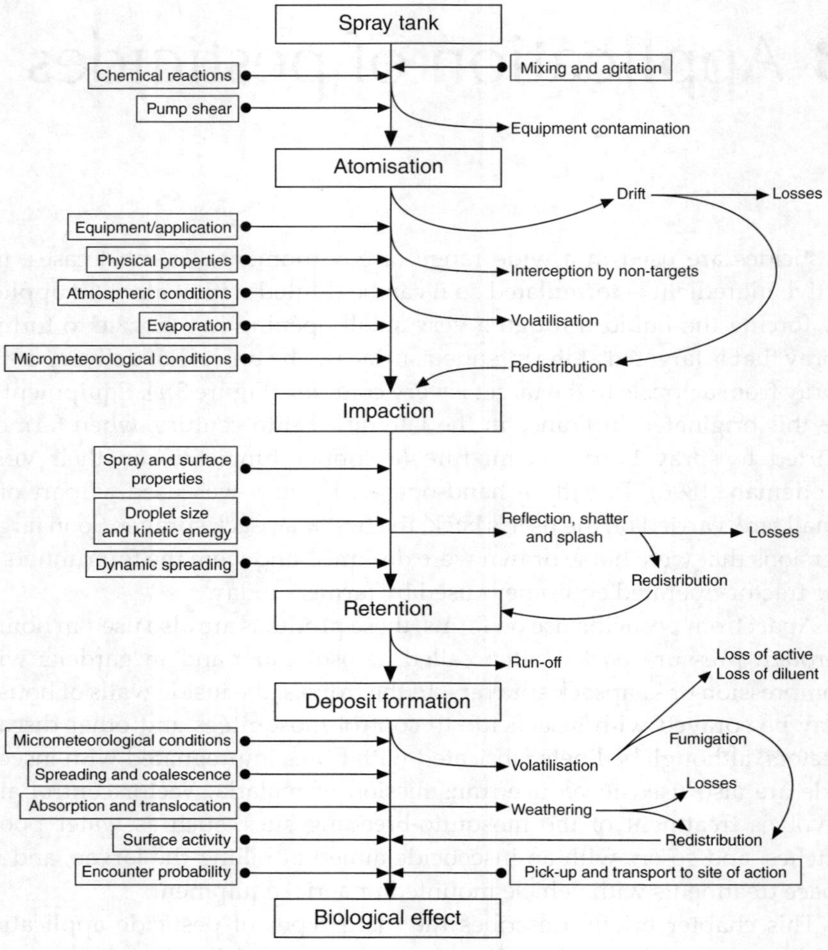

Fig. 3.1 The dose-transfer process, showing the complexity of effects on movement of the pesticide from the sprayer to achieve a biological effect (Matthews et al., 2014. Reproduced with permission of John Wiley & Sons).

(Matthews, 1999). Aerial spraying (Fig. 3.5) has been banned in Europe, due in part to increased risk of spray drift as spray is released from a greater height above the crop than with ground equipment. However, aerial spraying is permitted if a derogation is obtained. This has been mainly for applications to control pests in forests, for bracken and vector control. Elsewhere aerial application is still used on extensive crop areas. The hydraulic nozzles have previously been used, but many aircraft now have rotary atomisers to control droplet size more effectively. There is now greater interest in using unmanned aerial vehicles (UAV) (Fig. 3.5d), often referred to as drones, as the modern computer systems allow accurate control of the UAV over the crop. Unmanned helicopters were introduced initially in Japan to

(a)

(b)

(c)

(d)

Fig. 3.2 Tractor sprayer (a) mounted on three-point linkage (Photographs courtesy of Hardi International); (b) trailed (Photographs courtesy of Hardi International); (c, d) self-propelled (Photographs courtesy of Househam Sprayers)

(e)

(f)

Fig. 3.2 (*Continued*) (e) turf sprayer (Photographs courtesy of Hardi International) and (f) vertical boom sprayer in a glasshouse (Photograph by Richard Glass). (*See insert for color representation of the figure.*)

treat rice fields, but other countries are now doing trials, applying very low volumes of spray (Giles and Billing, 2014; He et al., 2014). A UAV can also be used for crop monitoring.

The spray produced will depend on the design of the nozzle, its shape and size as well as the pressure at which it is operated. Traditionally the pesticide was diluted in a large volume of water and often 1000 l or more were applied per hectare to crops. Once the exposed foliage was wetted, much of the liquid dripped to the ground and was wasted. This technique is still used in some parts of the world, but – with farm size increasing and a shrinking labour force – the need to enhance field work rates, the cost of collecting and transporting sufficient water to fields and the increasing

Fig. 3.3 Air-assisted boom sprayer provides a downwardly directed airflow of air into the crop.

(a)

Fig. 3.4 Air-assisted sprayers for bush and tree crops. (a) Axial fan in apples (Photograph by Graham Matthews);

(b)

(c)

Fig. 3.4 (*Continued*) (b) in vines (Hardi International); and (c) three-row Munckhof air system (Photograph by Jerry Cross). (*See insert for color representation of the figure.*)

(a)

(b)

Fig. 3.5 Aerial spraying (a) fixed wing aircraft, (b) helicopter spraying, (c) GPS tracking for more accurate treatment and record exact where the spray was applied and (d) UAV spraying a crop (Photograph by K. Giles, UC Davis, USA). (*See insert for color representation of the figure.*)

(c)

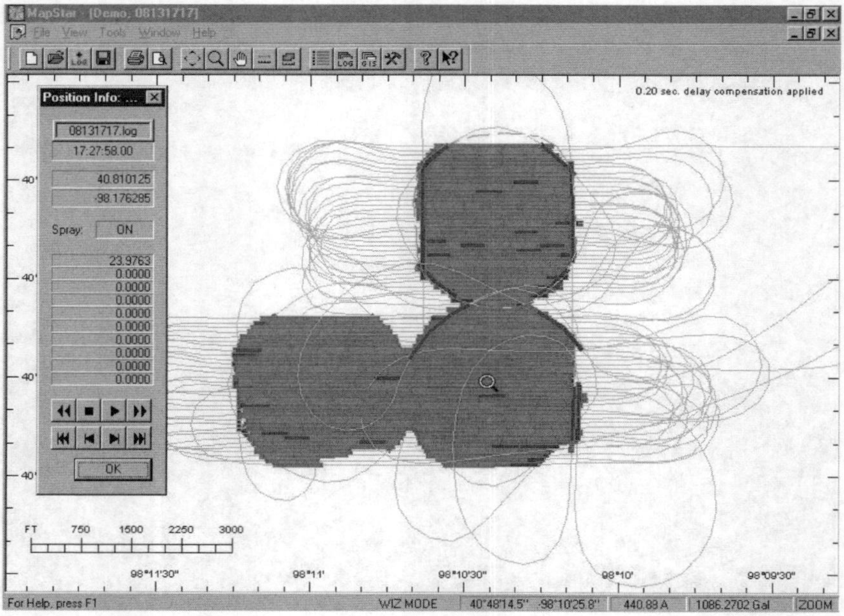

(d)

Fig. 3.5 (*Continued*)

recognition of the wastage of chemicals with high volume spraying have led to the application of much lower spray volumes. In the UK, the recent trend has been to reduce sprays on large-scale farms from over 200 l per hectare to between 80 and 120 l per hectare. At the same time the speed of travel across fields is increasing. Boom widths have increased with 18–24 m booms being preferred to increase work rates; with even wider booms (e.g. 36 m) where the land is very flat. Choice of boom width is usually dictated as a multiple of the seeder width, thus with a 4 m seeder, a 20 m boom would be selected, to accommodate 'tramlines' that allow tractor access at all stages of crop growth.

Apart from the tractor-mounted, trailed or self-propelled boom sprayers used in arable farming, there are downwardly directed air-assisted sprayers for treating field crops (Fig. 3.3). Interest in air-assisted sprayers for treating arable crops has increased where it is important to reduce spray drift. Air is ducted through a sleeve to create an air-curtain that propels droplets downwards across the width of a spray boom. In this way, there is better penetration of crop canopies and spray droplets are less prone to downwind movement across the top of the crop canopy. However, the foliage has to filter the droplets projected into the canopy, otherwise there may be more drift if the air bounces back from the ground. Air-assisted sprayers have been developed primarily for treating orchards (Fig. 3.4). Various designs are used with axial, centrifugal and cross-flow fans to move air with droplets into crop canopies. Where relatively low trees are planted improved spray distribution has been achieved using multiple-row sprayers with nozzles mounted in the inter-row space (Wenneker et al., 2014). Some dwarf-tree orchard crops are now treated with tunnel sprayers (Doruchowski and Holownicki, 2000) in which a proportion of the spray collected on the tunnel is recycled, saving the farmer about 30% of the spray (Jamar et al., 2010). Vertical booms on a small motorised sprayer (Fig. 3.2f) have now been introduced in glasshouses to replace manual spraying with a lance, a technique that exposes the operator to the spray.

Hydraulic nozzles

Typically nozzles have been designed to provide either a cone or flat fan-shaped sheet of spray. A number of different flat-fan nozzles are now available (Fig. 3.6). These have always been preferred on large tractor equipment, where the nozzles are mounted across a horizontal boom. Cone nozzles in contrast have been used on hand-operated sprayers and for equipment used in orchards in which spray droplets are projected in an airstream into the crop.

Nozzles were manufactured using brass, but as this metal is 'soft', the orifice can soon be eroded by particles of sand or other debris, suspended

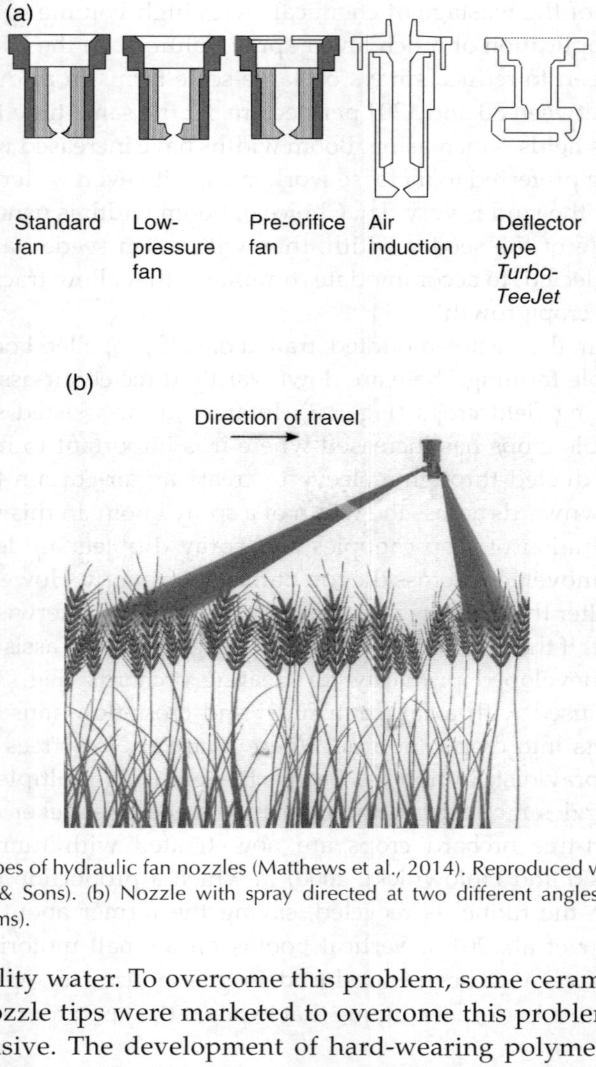

Fig. 3.6 (a) Types of hydraulic fan nozzles (Matthews et al., 2014). Reproduced with permission of John Wiley & Sons). (b) Nozzle with spray directed at two different angles into the crop (Spraying Systems).

in poor quality water. To overcome this problem, some ceramic or stainless steel nozzle tips were marketed to overcome this problem but were more expensive. The development of hard-wearing polymers has enabled farmers to use moulded plastic nozzles rather than the more traditional machined designs. This change in manufacturing technique provides nozzle tips of consistently high quality at a lower price, but inevitably erosion of the orifice will occur. Less-expensive ceramic tips are also now available due to a new manufacturing technique, and these are resistant to erosion. Nozzle tips are readily changed since they can be fitted using a screw or bayonet retaining cap at each mounting point or in matched sets across the boom to meet a range of contrasting application needs.

Flat-fan nozzle tips are now colour-coded to an international standard so farmers are able to select a set of the same output. The colour does not indicate the spray angle or type of tip. A filter, often of 50 mesh is fitted in the nozzle body to prevent blockages of the small orifice while spraying. A check valve is used at the nozzle to prevent liquid dripping from the lance or boom.

Farmers can spend large amounts on pesticides and often forget that by investing in new nozzle tips they can save money. A farmer can spend between £10–50 per hectare for the pesticide products, so if we assume £21 per hectare, including mixing an adjuvant, the total cost of chemical is £10 500 with 500 ha. If the nozzle is eroded and the flow rate has increased by 5%, the farmer could spend £11 025, an extra £525, whereas the cost of new nozzle tips across the boom would be less than £75.

Until the development of laser equipment to measure the size of the spray droplets from different nozzles, the main criterion was the output of the nozzle and the shape of the spray pattern. However, detailed assessment of the spray spectra confirmed that hydraulic nozzles produce droplets of a very wide range of sizes, some extremely small are less than 100 micrometres (µm) in diameter, while other droplets exceed 500 µm. This range of droplet sizes varies with nozzle design and use and can – with some pesticides – have a major effect on efficacy, crop selectivity and losses of spray beyond the intended treatment area. Drift losses are not just recent issues. After some hot weather in the UK in May 1976, concern increased that herbicides, which were applied to cereals, were adversely affecting vegetable crops downwind of treated fields. Investigations revealed that apart from volatile spray deposits, being carried downwind, there was a risk that the smallest spray droplets were also liable to be carried by air movement out of the treated fields. In consequence the British Crop Protection Council's Working Group on Chemical Applications initiated a study of droplet spectra from different nozzles. This led to a spray classification scheme (Fig. 3.7) in which different sprays could be classified as 'fine', 'medium', or 'coarse' (Doble et al., 1985). Nozzles that produced a high proportion of droplets in a 'very fine' category were not recommended for treating fields, because of the hazard of drift, although they could be used in glasshouses or other indoor situations, for which, subsequent studies have separated 'mists' and 'fogs' within the 'very fine' category (Matthews and Bateman, 2004). At the other extreme 'very coarse' sprays were generally not recommended as the large droplets tend to bounce off foliage, as leaves tend to have a waxy surface. This system of classification was modified to include a drift potential factor (Southcombe et al., 1997) to define more exactly the boundaries between the different categories. With further modifications this system of classification has been adopted in Europe and the USA. In the American standard, 'extra coarse' is an additional category. Nozzle manufacturers have included information on spray quality in their catalogues and other data are available on web sites such as www.dropdata.net. Most farmers would use a 'medium' spray, but even with these nozzles, some of the spray droplets can be carried downwind.

Apart from selection of nozzles that produce fewer small droplets, to reduce drift, unsprayed buffer zones have been introduced (see Chapters 5 and 6).

In an effort to improve spraying, nozzles have been designed in which air is mixed with the spray liquid. In some cases air is forced into the nozzle

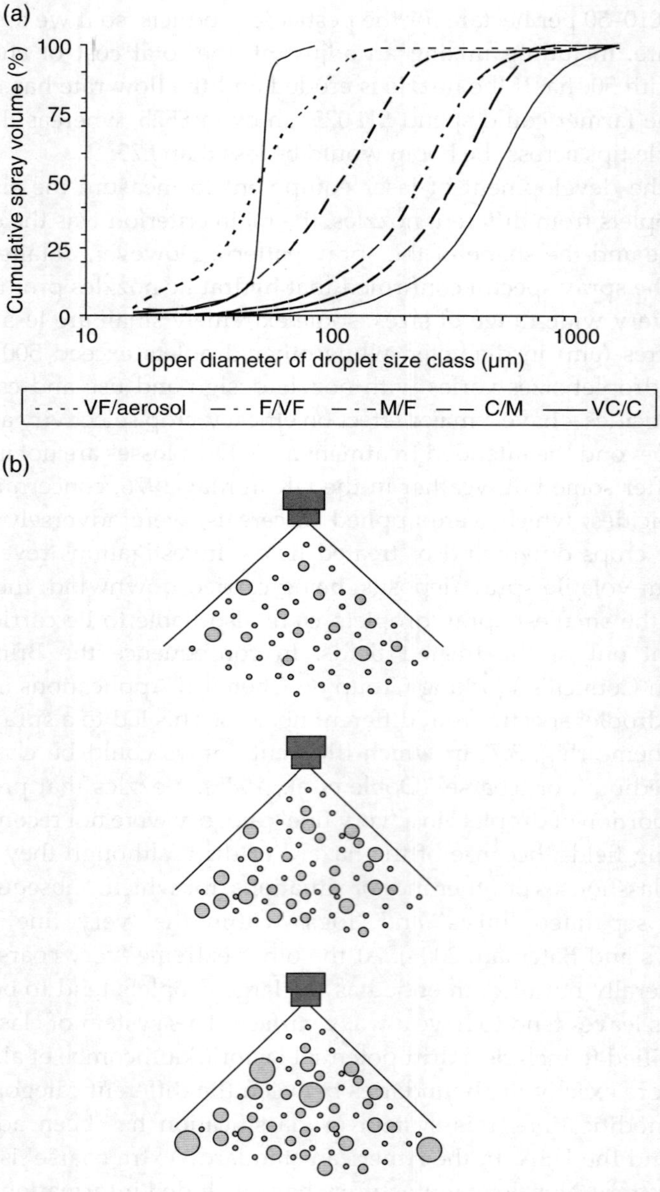

Fig. 3.7 (a) BCPC spray classification with hydraulic nozzles. VF, very fine; F, fine; M, medium; C, coarse; and VC, very coarse spray. The more vertical line is for a rotary atomiser. The position of this line depends on atomiser speed and flow rate. (b) Diagram showing range of droplet sizes within a fine, medium and coarse spray.

using a compressor while in others air is sucked into the nozzle using a venturi. 'Twin-fluid' nozzles in which air is mixed with the spray liquid inside the nozzle have been developed to cope with different tractor speeds (from 6 to 20 km/h) and thus different flow rates, while maintaining a similar droplet size range (Combellack and Miller, 1999). Spray quality can be adjusted in the tractor cab without changing nozzles. This allows a very coarse spray to be applied adjacent to sensitive areas, such as a watercourse, and then change quickly to a medium or fine spray to optimise spray deposition over the remainder of the field.

Nozzles which suck in the air, known as air induction (AI) nozzles (example in Fig. 3.6), have become popular in Europe and elsewhere, especially those given a 3* rating indicating >75% drift reduction. Several different manufacturers have produced AI nozzles with different coarse spray spectra (Piggott and Matthews, 1999). To understand differences between these nozzles Butler Ellis et al. (2002) investigated design parameters. Increasing both the Venturi throat diameter and final orifice size increased airflow rate, but compared to the final orifice size, air intake did not markedly affect droplet size. However, the proportion of very small droplets from AI fan nozzles is significantly less than with standard fan nozzles (Fig. 3.8), so the risk of downwind drift is reduced. With air entrapped inside them, the larger droplets are then less likely to bounce off foliage; the large – but slower – droplet is 'cushioned' at impact. However, the efficacy of some herbicides has been eroded by the more extensive use of low-pressure nozzles producing larger drop sizes, especially where the target surface is a vertical stem, e.g. grass weed. To overcome this problem, nozzles can be angled to project spray more directly at the target surface. Some nozzle tips have two separate orifices, usually arranged so that one sprays forwards and the other backwards, while some may have quite different trajectories of spray to target upper and lower sections of plants (Fig. 3.6b).

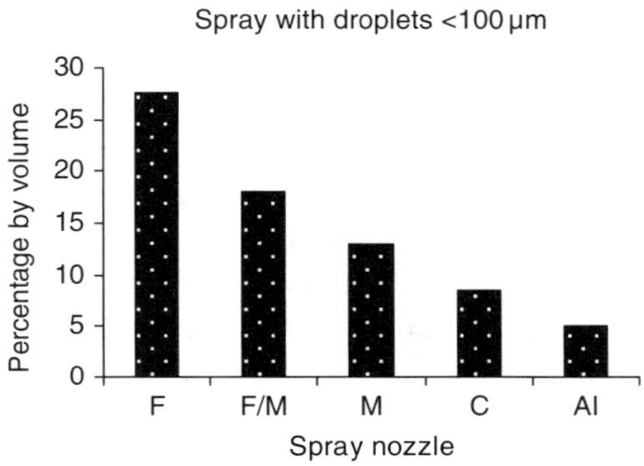

Fig. 3.8 Percentage by volume of droplets smaller than 100 μm, showing least spray in small droplets with air induction (AI) nozzles.

Field experiments have shown that variations in boom height, nozzle size, forward speed and nozzle operating pressure all affect the potential drift. The scale of drift risk is further compounded by the characteristics of the surface over which drift is measured as surface friction and meteorological variables all influence the movement of droplets while airborne (Bache and Johnstone, 1992). Models have been developed to predict the potential extent of spray drift. In the USA, the EPA uses the AgDISP model to assess potential drift, but changes in meteorological factors will influence how far droplets will travel. Nevertheless the models provide some comparative data to improve recommendations on nozzle selection in relation to drift (Holterman et al., 1997), by using some comparative data on nozzle selection. Some models have under-predicted spray drift (Phillips and Miller, 1999), while other studies have shown that the position of nozzles on the boom and their spacing also influences drift potential (Murphy et al., 2000). In the USA, the adoption rates of drift-reducing practices by commercial applicators has been highly variable (Reimer and Prokopy, 2012), which may be why the EPA has now introduced a Drift Reduction Technology (DRT) programme. The aim is to encourage the manufacture, marketing and use of spray technologies scientifically verified to significantly reduce pesticide drift. The focus will initially be on spray technologies used on row and field crops by ground or aerial application equipment nozzles, spray shields and drift-reducing adjuvant chemicals, but will ultimately include technologies for orchard and vineyard crops. Under this programme the DRT will be given a star rating of spray drift:

- One star – 25–49% reduction
- Two stars – 50–74% reduction
- Three stars – 75–89% reduction
- Four stars – 90+% reduction

(http://www2.epa.gov/reducing-pesticide-drift/about-drift-reduction-technology-program). (See also Chapter 6.)

While air induction nozzles reduce drift, coverage is not so good on certain targets such a grass weeds like black grass seedlings prior to the three-leaf stage (Powell et al., 2003). Adaptations by angling an air induction nozzle have subsequently been developed, for example to improve efficacy of late fungicide sprays on cereals (Robinson et al., 2003). However for improved coverage with small droplets an external air-jet directed at the output of a fan nozzle causes improved break-up of the spray and entrains the small droplets within the air-stream, thus achieving less drift with a finer spray (Matthews and Thomas, 2000a, 2000b).

In row crops it is possible to spray a narrow band of pesticide either along the crop row or inter-row, depending on requirements. This is possible with 'even-spray' fan nozzles. On some sprayers the nozzle is operated within a cover to avoid any spray drift. In conjunction with GPS treatments can be applied very accurately. Their use is increasing in precision farming, as good weed management is possible within limited space between rows. The use of herbicides will also increase with genetically modified crops, which allow certain herbicides to be sprayed over the crop. Confining the pesticide to a narrow strip not only reduces the cost of treatment, but it allows integrating of other techniques such as inter-row hoeing.

International standards for hydraulic nozzles enable farmers to select the most appropriate nozzle for a particular pest problem. Colour coding of flat-fan nozzles shows what their 'throughput' is at a specified pressure. All makes of nozzle should also fit the standard size nozzle body. Nozzle bodies do vary in terms of their fitting to the boom and type of retaining cap (bayonet or screw). Some booms have nozzle turrets so that different types of nozzle can be selected in the field. On advanced equipment, with modern control equipment in the cab (Fig. 3.9), each nozzle has a solenoid valve to enable the nozzle to be computer controlled, spraying sections of a field in relation to GIS/GPS data, for example when spraying patches of weeds rather than the whole field. Selection of a nozzle involves deciding the spray pattern (fan, cone), spray angle (e.g. 80°), output (l/min) and spray quality at the desired operating pressure. The sprayer must also be calibrated to check the output is that required at the forward speed of the sprayer. In precision agriculture,

Fig. 3.9 Modern control equipment in tractor cab (Photograph courtesy of Househam Sprayers).

some rotary atomisers and hydraulic nozzles are mounted under a guard to apply a pesticide to a specific narrow swath without spray drift (Fig. 3.10a). The reduction in drift (Fig. 3.10b) makes the equipment ideally suited for treating the inter-row without affecting the crop (Clayton, 2014).

(a)

(b)

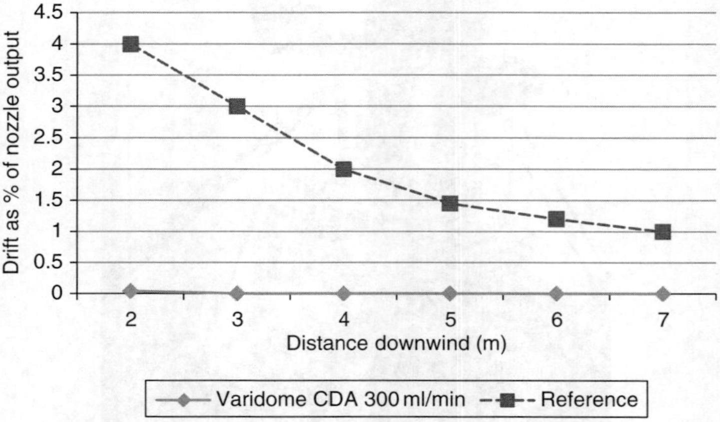

Fig. 3.10 (a) Nozzles protected by a shield to exclude downwind drift of spray for precision application of herbicides Varidome (Photograph courtesy of Micron). (*See insert for color representation of the figure.*) (b) Percentage drift from Varidome CDA shield compared to the reference nozzle (dotted line).

Sprayer testing

In Europe, sprayer testing is now mandatory to ensure farmers maintain their equipment to a minimum standard, as required by the European Directive 2009/128/EC on the sustainable use of pesticides, that requires protection of the environment. By keeping sprayers well maintained, the aim is to avoid overdosing, leakages from equipment and spillage of pesticides while treating crops. The check is equivalent to the routine inspection of vehicles, required before a vehicle can be taxed. As discussed in Chapter 2, the National Sprayer Testing Scheme is responsible for these tests in the UK.

Rotary atomisers

A major criticism of hydraulic nozzles is the wide range of sizes that are produced. A narrower droplets spectrum can be achieved by using a rotary atomiser or spinning disc. With these the average droplet size is dependent on the speed at which the outer surface of the rotating nozzle travels. Higher speeds produce smaller droplets. Initially these were largely confined to laboratory studies, but where ultra-low volumes (ULVs) (<51 per hectare) and very low volumes (VLVs) are applied, the rotary atomiser is usually better as the orifice in the flow constrictor is less likely to block. Equipment with rotary atomisers/spinning discs has been used mainly in arid areas, to treat

(a)

Fig. 3.11 Rotary atomiser CDA spinning disc sprayer (a) insecticide application; (b) applying herbicide (Photograph courtesy of Micron); and (c) 'Electrafan' sprayer being used in a glasshouse (Photograph courtesy of Micron). (*See insert for color representation of the figure.*)

(b)

(c)

Fig. 3.11 (*Continued*)

cotton, notably in West Africa, cowpea and other crops (Fig. 3.11a) and for locust control, and on aircraft (Fig. 3.5). Some rotary atomisers are fitted to small sprayers with an engine driven fan to produce a mist (Fig. 3.11c) that settles on foliage and provide a more residual effect than a space treatment described below. With slow rotational speeds, usually around 2000 rpm, small hand-carried spinning-disc sprayers have also been used for herbicide treatments in amenity areas and in forests, as droplets are consistently large enough to minimise spray drift (Fig. 3.11b).

Compression sprayers

These are popular for garden use. Most have a small plastic tank (<10 l) and a hand-operated pump which also provides the lid (Fig. 3.12a). A small lance has usually been fitted with a trigger valve and an adjustable cone nozzle, although these should not be used as manual adjustments to the spray angle exposes the hand to pesticide. Settings can vary from a straight jet to a wide cone, the latter having the finest spray while the pressure is high. Spray quality will vary depending on the pressure and the way the nozzle is adjusted. Apart from the extreme positions of the nozzle, seldom can it be used consistently at intermediate settings. A compression sprayer tank is not completely filled, so there is an airspace pressurised by pumping. As spray is applied, the pressure in the tank decreases and so users have to stop after a while and re-pressurise the tank. Ideally a control flow valve is fitted to ensure uniform pressure at the nozzle. Compression sprayers that meet the WHO specification have been recommended for indoor

(a)

Fig. 3.12 (a) Small compression sprayer (Photograph by Graham Matthews).

(b)

(c)

Fig. 3.12 (*Continued*) (b) In warehouse (Killgerm Chemicals). (c) Compression sprayer with electric pump (Photograph courtesy of Gloria). (*See insert for color representation of the figure.*)

residual spraying to control disease vectors such as mosquitoes (Matthews, 2011). They have also been used in warehouses to provide a residual spray to control storage pests (Fig. 3.12b). Some of the smaller compression sprayers now have a battery powered pump to eliminate the manual pumping (Fig. 3.12c).

Knapsack sprayers

The lever-operated knapsack sprayer (Fig. 3.13a–c) requires continual pumping in contrast to the compression sprayer and is designed to pump the spray liquid, rather than air. It is probably the most widely used type of sprayer, being suitable for small farms and all areas inaccessible to vehicle equipment. Design of these sprayers has improved with modern manufacturing techniques and efforts to meet international standards, such as the minimum requirements for these sprayers published by FAO (Anon, 2001). International standards require a large tank opening with a deep set filter to facilitate filling, a lid with valve to air to enter but prevent liquid to escape, a minimum of 0.5 m lance with a trigger valve incorporating a filter and parking position to protect the nozzle when not in use. Traditionally the tank is fitted with only two shoulder straps, but ideally there should also be a waist strap to ensure the tank is tight on the operator's back to enable efficient transfer of energy used in operating the lever to the pump. The end

(a)

Fig. 3.13 (a) Lever-operated knapsack sprayer used in rice (Photograph by Graham Matthews).

(b)

(c)

Fig. 3.13 (*Continued*) (b) Spraying pavement to control weeds; (c) with shield around nozzle to protect spray from wind.

(d)

(e)

Fig. 3.13 *(Continued)* (d) Knapsack with electric pump to eliminate manual pumping (Photograph courtesy of Birchmeier). *(See insert for color representation of the figure.)* (e) Knapsack motorised mistblower (Photograph by Graham Matthews).

of the lance should allow any type of hydraulic nozzle to be fitted so the spray output and pattern meets the specific requirements for different pesticides and crops.

Although the majority of these knapsack sprayers have a manually operated pump, some have a pump driven by an electric motor with a rechargeable battery (Fig. 3.13d). Some have a petrol-driven engine (usually two-stroke engine) can operate at higher pressures. Some engine-driven knapsack sprayers also have a fan to provide an air-stream to project the spray into trees and other crops. These are referred to as knapsack mistblowers, and are fitted with an air-shear nozzle or in some cases, a rotary atomiser. Knapsack mistblowers (Fig. 3.13e) tend to apply lower volumes, and thus use sprays with a higher concentration of pesticide.

Home and garden use

Traditionally some pesticides, especially insecticides, are sold as low concentration dusts in 'puffer' packs. This allows small areas, such as around doorways and under sinks where cockroaches or ants occur, can be treated. The amount of dust emitted from individual puffs can vary depending on the amount in the container, the angle at which it is used and severity of the squeezing of the container. In many situations application of 'gels' or other techniques of pest management have replaced dusts.

Pressure packs, often referred to as aerosol cans, are a very popular means of spraying small quantities of pesticide. The chemical is dissolved in a solvent and sealed in a robust container with a propellant (compressed or liquefied gas). Fluorinated hydrocarbons are no longer used and have largely been replaced by compressed air or butane. Operating the valve in the top of the can allows the pressure within the can to force liquid up a dip-tube and through the valve, which is also designed as a nozzle. Propellant reaching the atmosphere causes the liquid pesticide to break up into droplets. Depending on the volatility of the solvent and other factors, the droplet size produced can vary, but for insecticide application tends to be less than 30 μm.

A less-expensive method of producing an aerosol with a hand-operated pump is by using a Flit gun. The small container can be refilled with pesticide. A small air pump, similar to a bicycle pump, is used to create a jet of air across the top of a dip tube to atomise the liquid and propel the droplets into the atmosphere.

Pre-diluted pesticides may also be sold in plastic containers with a trigger-operated cone nozzle (Fig. 3.13). These ready-to-use (RTU) sprayers are very useful, where only small area, such as a few rose bushes, need treatment.

Weed wipers

When there is a risk of spray drift even over a very short distance to a sensitive plant, it has been possible to apply a translocated broad-spectrum herbicide, such as glyphosate with an absorbent sponge-like material, the 'wick' which is attached to a reservoir of the herbicide (Fig. 3.14). Weed wipers need to be designed so that the wick is sufficiently wet to transfer chemical to the weed, but not so wet that the herbicide drips

Fig. 3.14 Ready-to-use (RTU) spray for garden use (Photograph by Graham Matthews).

from it. By touching surfaces, the wick can get dirty and affect chemical transfer. Equipment can be mounted on a tractor or be hand-carried. They have been used to control specific weeds in set-aside and managed buffer zones.

Space treatment equipment

Specialised equipment is used to apply small droplets inside warehouses, glasshouses and other spaces where flying insects need to be controlled (Fig. 3.15). The same equipment is occasionally used for applying certain fungicides, and also for treating outdoor areas where adult mosquitoes and other disease vectors require control. Small droplets are used so that they remain suspended in air as long as possible, although in still air even the smallest droplets will gradually fall by gravity over several hours on to exposed horizontal surfaces.

The use of thermal foggers was the main method of producing droplets smaller than 25 µm by vaporising the liquid containing the insecticide at about 400–500°C and allowing the vapour to condense as a dense white cloud of very small droplets. The insecticide was normally mixed with odourless kerosene or equivalent, but the trend is towards water-miscible

Fig. 3.15 Weed wiper (Photograph by Graham Matthews).

(a)

Fig. 3.16 Space treatments: (a) thermal fogging in a glasshouse, (b) in a plantation (Photograph by Graham Matthews), (c) vector control and (d) vehicle mounted cold fogger (Photograph courtesy of Micron). (*See insert for color representation of the figure.*)

products that have to be mixed with a 'carrier'. Small hand-carried thermal foggers generally have a pulse-jet engine (Fig. 3.16a–c). Rather than use a high temperature, cold foggers use a vortex of air to shatter the spray into small droplets. Most vehicle-mounted cold foggers have a 16–20 hp engine,

(b)

(c)

Fig. 3.16 *(Continued)*

to drive a blower to provide low-pressure air to a vortical nozzle (Fig. 3.16d). Other cold fog equipment, which is electrically operated, can be set up with a timer to treat a space, such as a glasshouse, when not occupied. In contrast thermal fogging equipment must not be left unattended due to a fire risk. Insecticides formulated for cold fogs are increasingly diluted in water but some formulations incorporate a chemical, which forms a film on the

(d)

Fig. 3.16 (*Continued*)

surface of droplets to reduce evaporation and thus prevent droplets shrinking and becoming too small.

Small droplets in glasshouses will be sucked out through gaps in the structure due to wind over the structure. However if the building is well constructed, a fog can remain effective for several hours, before vents are opened to ventilate the area before access is allowed. Respiratory protective equipment (RPE) has to be worn, when operating fogging equipment inside buildings, such as warehouses and glasshouses.

Granule application

A few pesticides are formulated as dry granules and used directly by incorporation into soil to control pests, such as potato cyst nematode, or spread over the ground or crop to control slugs or weeds. Granules are less easy to meter accurately compared with pumping a liquid through nozzles. Specialised equipment has been developed, which can incorporate closed transfer of the product directly to the dispensing hopper and specific metering units. Manually operated granule applicators have been used to apply nematicides on bananas.

Some of the most toxic nematicides, considered too hazardous to apply as sprays, were recommended for application as granules, but some of these, for example aldicarb, have also been withdrawn. Fosthiazate is an organophosphate (WHO Class II) which recommended and is packaged in a Surefill container for closed transfer of the product directly to the dispensing hopper on a granule applicator. Using a nematicide bed tiller is the most effective way of applying nematicide for crops such as potatoes to control potato cyst nematodes. Care is needed to ensure the toxic granules are not left on the soil surface to avoid exposing birds to them, so all applicators have to be fitted with an in-cab device allowing operators to shut off nematicide flow at least 3 m from the end of each row.

In seeking less-toxic bio-pesticides, the known effect of naturally occurring soil bacteria *Pasteuria* spp can now be exploited due to a revolutionary in vitro production process, which has led to development of cost-effective nematicides with a novel mode of action. The first of these is being released to control soybean cyst nematode. Apart from granular application, formulations containing *Pasteuria* can be applied via standard ground spray equipment or through chemigation.

Seed treatment

This is an increasingly important method of pesticide use, normally as a systemic fungicide or insecticide aimed at protecting the young seedlings, but sometimes with a safener to protect the seedlings from the effects of a herbicide. Seeds may be pelleted and provided with some nutrients as well. Localised treatment reduces the dose applied. Prophylactic treatment is generally cost effective where soil pathogens may drastically reduce seedling survival. Any early insect attack by sucking pests can also be controlled when a spray would be very inefficient due to the small size of the seedling relative the ground area. Some seed may remain on the soil surface after drilling. De Snoo and Luttik (2004) estimated from field data that for risk assessment, the percentage of seed remaining on the soil surface was 3.3% for standard drilling in the spring, and 9.2% in the autumn, but this was reduced to 0.5% with precision drilling.

The widespread use of neonicotinoid insecticides as a seed treatment significantly reduced the need for spray applications, but concern about the decline of bees, known as colony collapse disorder, led to a moratorium on their use in Europe as it was considered that the neonicotinoids were one of the factors causing bee mortality. Similar concerns have been now expressed in North America. However, bee mortality was also caused by the *Varroa* mite and viruses transmitted by mites. With a very low dosage applied to seeds, it is questionable whether the amount of insecticide in flowers of crops, such as oil seed rape, later in the season would have caused high mortality. Where an acute problem undoubtedly occurred was due to dust from seed treatments being vented from equipment used in planting, which was airborne and would have had a direct effect of bees in the same locality. This particular problem was overcome by modifications to the formulation of the insecticide used on seeds and to the seeder equipment to prevent release of dust into the environment. It is possible that the presence of neonicotinoids in bees could have had a synergistic effect on infectious agents such as *Nosema* which resulted in colony collapse. In contrast in Australia where *Varroa* mites are not present, bee populations have thrived, despite the use of neonicotinoid insecticides (see also Chapters 6 and 8).

Storage of pesticides and equipment

It is essential that pesticides are stored safely as the concentrated formulations pose the most risk to human health and the environment. Equipment is best stored separately away from chemicals. In the UK commercial pesticide stores have to be inspected annually to ensure that the building is soundly constructed with fire resistant materials, well-lit and ventilated on a suitable site with adequate capacity and segregation of products. Never store pesticides in places where flooding is possible or in places where they might spill or leak into wells, drains, ground water or surface water. Pesticide stores should be bunded and have a sump to prevent spillages reaching watercourses. This is also important if a fire occurs and the water is used to quench the flames.

The building must be dry and frost-free with appropriate warning signs and secure against theft and vandalism. Suitable access and exits must also be provided with provision to contain any spillage or leakages. Staff must be trained. The following guidelines also need to be followed where pesticides are stored on farms:

- Avoid excessive quantities in stock by having only the amount needed in the near future.
- Keep all pesticides in a locked, ideally separate store or cabinet in a well-ventilated utility area, barn or garden shed, and ensure that any spillages do not seep into the ground or enter watercourses. Where small quantities are involved, the locked cabinet must be high enough to out of reach of children. Never store pesticides in the same area as food, animal feed or medical supplies. In some countries, where pesticides were in the same general stores, legislation now insists pesticides are sold from separate shops.
- Pesticides must be stored in their original containers, with the label listing ingredients, directions for use, and first aid steps in case of accidental poisoning. Follow all storage instructions on the pesticide label. *Never* transfer pesticides to soft drink bottles or other containers. Children or others may mistake them for something to eat or drink.

- In domestic use always use containers that are child-resistant and close the container tightly after using the product. However, child-resistant does not mean child proof, so extra care is needed to store the container properly in a locked cabinet as described above.
- If the contents of the container cannot be identified, or how old the contents are, follow advice on safe disposal.
- Further information on storage is available at http://www.hse.gov.uk/pubns/ais16.pdf, http://www.epa.gov/pesticides/regulating/store.htm, http://entweb.clemson.edu/pesticid/saftyed/storage.htm, http://www.environment-agency.gov.uk/netregs and http://www.agregister.co.uk/product
- On the http://www.hse.gov.uk/pubns web page, there are publications, including those concerning record keeping and amenity use.

Timing and number of spray applications

In some crops, especially in the developing countries, recommendations were simplified by recommending that sprays were repeated at weekly, 10-day or longer interval on a regular calendar schedule. This meant that some applications might be made when no pest or disease was present, so the trend with integrated pest management has been to recommend crop monitoring and only apply a pesticide if the pest or disease economic threshold was exceeded. That is, if pest numbers continue to increase, loss of crop yield and/or quality will reduce the farmer's income unless appropriate control measures are undertaken.

Scouting

Crop monitoring can be done in various ways depending on the pest and crop.

Some farmers employ specialist consultants to 'walk' their crops and decide whether a pesticide is needed. In some crops routine scouting is essential to assess if an insect population needs control. Such scouting can be assisted by using pheromone or various types of sticky trap to sample the pest population. While traps may indicate the presence of a pest, the scout may still need to examine the crop as trap catches may not be directly correlated with the pest population within the crop. Where a disease, such as potato blight is anticipated, mini-meteorological stations close to the crop can monitor temperature, humidity and rainfall so that the farmer can assess with assistance of decision support systems, whether conditions will favour the disease.

Some pre-emergence herbicides could be applied before weeds were present, often at the time a crop was sown, but the trend has been to apply selective herbicides when crop walking indicates weeds are present in sufficient numbers to justify treatment. With herbicide-tolerant, GM crops, a broad-spectrum herbicide can be applied later in the season. Small cloches can be used to accelerate germination or growth of latent weeds to aid identification and likely control measures.

Generally economics dictate the minimum number of applications that are applied, but some crops may require several different pesticides over a period of several weeks. The development of insect-resistant GM crops, such as Bt cotton will reduce the number of spray applications, typically needed on the crop for a particular pest. Some pesticides are applied as late sprays to protect the harvested produce during storage. A seed treatment is a technique that allows a low dose of pesticide to be applied that will protect the young plants during crop establishment.

The ideal situation is when a pesticide is accurately timed and the dose minimised to just keep pest populations below the economic threshold. Optimising an application of herbicide, and other pesticides depends on timeliness, as the dose required for weeds or insects increases with their size or age. Delaying a herbicide spray by a short period, for example from three of four leaf stage can increase the dose required by 50% of more (Miller, 2003).

This dose should be applied only where it is needed in the crop. However, targeting the dose to the particular biological site, where action is needed is difficult and in reality the farmer generally has to spray the whole field, although at least for weeds research has indicated that spraying individual patches of weeds is possible. In some situations, a lower dose of insecticide has the advantage that it will be less harmful to beneficial insects in integrated pest management. Angling nozzles has helped to increase deposits on vertical target such as wheat stems.

In a crop, a spray will leave a pattern of droplets on the foliage and the ground under the plants, which will dry to form a surface deposit of the pesticide. If too high a volume is applied, the surplus liquid will drip from leaves and eventually increase the deposit on the soil. Although many have considered that high volume spraying more effective, as it is perceived to wet all surfaces, it can be very inefficient.

The deposit of pesticide active ingredient is then exposed to the effects of sunlight, rain and abrasion due to movement of foliage. Some active ingredients are readily absorbed into the plants and may move systemically up through the plants and accumulate on the upper leaves. Some may move only across a leaf – translaminar – while some, like glyphosate can be translocated downwards, thus in the case of grasses, the chemical reaches the rhizome. Deposits that remain on the leaf surface are effective by contact action or may be ingested by an insect, but remain the most exposed to degradation. Rain within two hours of a spray application could wash off a spray deposit, although the formulation is usually designed to stick the deposit on the foliage. Ultra-low volume oil-based sprays had the advantage of being more rain-fast. Very fine particles can adhere extremely well to surfaces. If the pesticide is too volatile, loss of deposits by vapour lifting from the crop can occur. In some cases limited vapour action is useful as pests not in direct contact with a spray deposit may be killed, but downwind movement of pesticide vapour will cause environmental damage.

The development of some pesticides has aimed at increasing persistence, thus the synthetic pyrethroids are more stable in sunlight than the natural pyrethrins. Nevertheless, the effectiveness of a deposit will decrease over time. One factor is the growth of the plants, which will increase the surface area of the foliage and thus dilute the impact of a deposit. Rapid plant growth in a crop like cotton in the tropics required a weekly spray treatment during a sustained bollworm infestation. A low dose applied more frequently was more effective than attempting to apply a higher dose less often.

One of the factors on the label is the pre-harvest interval (PHI). This takes into account the persistence of the pesticide, the recommended dose and environmental/climatic conditions to indicate the period over which a deposit will have decayed and no longer leave a residue above the MRL in the harvested produce. Thus if a pest infestation is considered serious close to harvest, great care is needed to observe the PHI and use a chemical that is considered suitable for applying close to harvest. This is a particular problem with crops, such as lettuce eaten raw, where a fungicide may be needed to avoid a fungal disease *Botrytis* developing on the lettuce head between harvesting and being displayed in the marketplace.

Concentrations of pesticide solutions may also have upper limits imposed for safety. As volume rates decrease, the typical spray concentration used has increased but regulators may specify an upper limit for some actives that are not to be exceeded. Fears of operator exposure and inhalation may increase with these changes and may need to be controlled.

This chapter has shown the wide range of equipment available for applying pesticides and the problems associated with determining when a pesticide should be applied.

References

Anon (2001) *Guidelines on Minimum Requirements for Agricultural Pesticide Application Equipment*. FAO, Rome.

Bache, D.H. and Johnstone, D.R. (1992) *Microclimate and Spray Dispersion*. Elis Horwood, Chichester.

Butler Ellis, M.C., Swan, T. Miller, P.C.H., Waddelow, S., Bradley, A. and Tuck, C.R. (2002) Design factors affecting spray characteristics and drift performance of air induction nozzles. *Biosystems Engineering* **82**, 289–296.

Clayton, J. (2014) The Varidome precision band sprayer for row crops. *Aspects of Applied Biology* **122**, 55–62.

Combellack, J.H. and Miller, P.C.H.(1999) A new twin fluid nozzle which shows promise for precision agriculture. *Proceedings of the Brighton Crop Protection Conference – Weeds*, Brighton. BCPC, Farnham, pp. 473–478.

De Snoo, G.R and Luttik, R. (2004) Availability of pesticide-treated seed in arable fields. *Pest Management Science* **60**, 501–506.

Doble, S.J., Matthews, G.A., Rutherford, I. and Southcombe, E.S.E. (1985) A system for classifying hydraulic nozzles and other atomisers into categories of spray quality. *Proceedings of the Brighton Crop Protection Conference – Weeds*, Brighton. BCPC, Farnham, pp. 1125–1133.

Doruchowski, G. and Holownicki, R. (2000) Environmentally friendly spray techniques for tree crops. *Crop Protection* **19**, 617–622.

Giles, D.K. and Billing, R. (2014) Unmanned aerial platforms for spraying: deployment and performance. *Aspects of Applied Biology* **122**, 63–69.

He, X., Liu, Y., Song, J., Zeng, A. and Zhang, J. (2014) Small unmanned aircraft application techniques and their impacts for chemical control in Asian rice fields. *Aspects of Applied Biology* **122**, 33–45.

Holterman, H.J., van de Zande, J.C., Porskamp, H.A.J. and Huijsmans, J.F.M. (1997) Modelling spray drift from boom sprayers. *Computers and Electronics in Agriculture* **19**, 1–22.

Jamar L., Mostade, O., Huyghebaert, B., Pigeon, O. and Lateur, M. (2010) Comparative performance of recycling tunnel and conventional sprayers using standard and drift-mitigating nozzles in dwarf apple orchards. *Crop Protection* **29**, 561–566.

Lodeman E.G. (1896) *The Spraying of Plants*. Macmillan, London.

Matthews, G.A. (1999) *Application of Pesticides to Crops*. IC Press, London.

Matthews, G.A. (2011) *Integrated Vector Management*. Wiley-Blackwell, Oxford.

Matthews, G.A. and Bateman, R.P. (2004) Classification criteria for fog and mist application of pesticides. *Aspects of Applied Biology* **71**, 55–60.

Matthews, G.A. and Thomas, N. (2000a) Working towards more efficient application of pesticides. *Pest Management Science* **56**, 974–976.

Matthews, G.A. and Thomas, N. (2000b) Effective use of air for low drift of fine sprays. *BCPC Symposium* **74**, 221–224.

Matthews, G.A., Bateman, R.P. and Miller, P.C.H. (2014) *Pesticide Application Methods*, 4th Edition. Wiley-Blackwell, Oxford.

Miller, P.C.H. (2003) The current and future role of application in improving pesticide use. *The BCPC International Congress*, Glasgow. BCPC, Farnham, pp. 247–254.

Murphy, S.D., Miller, P.C.H. and Parkin, C.S. (2000) The effect of boom section and nozzle configuration on the risk of spray drift. *Journal of Agricultural Engineering Research* **75**, 127–137.

Phillips, J.C. and Miller, P.C.H. (1999) Field and wind tunnel measurements of the airborne spray volume downwind of single flat-fan nozzles. *Journal of Agricultural Engineering Research* **72**, 161–170.

Piggott, S. and Matthews, G.A. (1999) Air induction nozzles: a solution to spray drift? *International Pest Control* **41**, 24–28.

Powell, E.S., Orson, J.H., Miller, P.C.H., Kudsk, P. and Mathiassen, S. (2003) Defining the size of target for air induction nozzles. *The BCPC International Congress*, Glasgow. BCPC, Farnham, pp. 267–272.

Reimer, A.P. and Prokopy, L.S. (2012) Environmental attitudes and drift reduction behavior among commercial pesticide applicators in a U.S. agricultural landscape. *Journal of Environmental Management* **113**, 361–369.

Robinson, T.H., Butler-Ellis, C.M.C. and Power, J.D. (2003) Evaluation of nozzles for the application of a late fungicide spray. *The BCPC International Congress*, Glasgow. BCPC, Farnham, pp. 273–278.

Southcombe, E.S.E., Miller, P.C.H., Ganzelmeier, H., Van de Zande, J.C., Miralles, A. and Hewitt, A.J. (1997) The International (BCPC) spray classification system, including a drift potential factor. *Proceedings of the Brighton Crop Protection Conference – Weeds*, Brighton. BCPC, Farnham, pp. 371–380.

Wenneker, M., Van de Zande, J.C., Stallinga, H., Michielsen, J.M.P.G., van Velde, P. and Nieuwenhuizen, A.T. (2014) Emission reduction in orchards by improved spray deposition and increased spray drift reduction of multiple row sprayers. *Aspects of Applied Biology* **122**, 195–202.

4 Operator exposure

Operators are involved in activities relating to the application of a plant protection products (PPPs), including mixing/loading the product into the application equipment, operating the application equipment and making repairs when necessary whilst it is being used, and emptying/cleaning the machinery/containers after use. Farmers or contract applicators can be employed for up to 8h while some home garden users may be applying a pesticide over a short period.

Exposure to pesticides tends to be greatest for those who mix and apply the sprays in the field, especially those who are employed by contractors or work on large estates and plantations where a pesticide may be applied on consecutive days and sometimes for a prolonged period during the year. Those preparing the spray are potentially at greatest risk of exposure to the concentrate pesticide product, whereas the applicator may only be exposed to the dilute spray. Worker exposure is therefore an important issue in occupational health and is an essential part of the risk assessments in pesticide registration (Van Hemmen and Brouwer, 1997). Hamey (2001) has pointed out the need to define representative use patterns in relation to interpret exposure assessments. Operators can be exposed to pesticide that reaches the skin (dermal), by inhalation and by accidental ingestion (oral), for example by eating while working. The most important of these routes is dermal for commonly used pesticide application techniques. The most important of these routes is dermal, so keeping the body well covered when using pesticides is so important (Fig. 4.1).

Decisions on operator exposure are based on a comparison of the no-observed-adverse-effect level (NOAEL) and an estimate of human exposure. The acceptable-operator-exposure level (AOEL) is derived from the NOAEL by dividing it by an assessment factor (usually by a factor of 100 which is essentially made up of two 10× uncertainty factors (Renwick, 1991; Anon, 1999a) to allow for inter- and intra-species variation as discussed in an earlier chapter. In some countries, such as the USA, they derive margin of exposure (MOE), or margin of safety, in a similar manner. As it would be impossible to measure the exposure in all situations, there is considerable reliance on experimental data obtained from particular usage situations

Pesticides: Health, Safety and the Environment, Second Edition. G.A. Matthews.

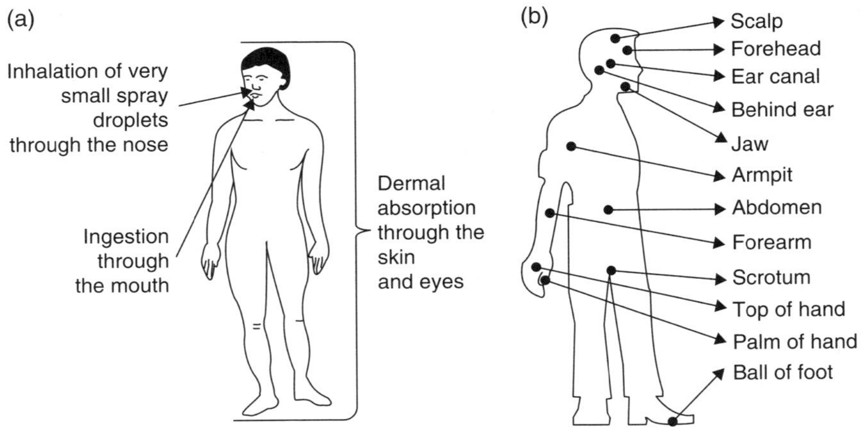

Fig. 4.1 (a) Routes of exposure to pesticides. (b) Absorption of pesticide in different parts of the body.

that have been incorporated into generic models such as the European Predictive Operator Exposure Model (EUROPOEM) (Van Hemmen, 2001). Field assessments were carried out especially in Southern Europe to provide more data in developing EUROPOEM (Machera et al., 2001; Glass et al., 2002b). The EUROPOEM database is available to be used in conjunction with existing models developed in the UK and Germany. Further studies (Gerritsen-Ebben et al., 2014) are aimed at refining the earlier models using outdated empirical data and develop a new framework integrating all available exposure data (see https://secure.fera.defra.gov.uk/browse). An initial examination of the BROWSE model indicated a higher exposure for mixing and loading the sprayer in contrast to some earlier models, but corresponded well with EUROPOEM. Ross et al. (2015) provide a review of models for estimating human exposure to pesticides.

In Europe under Regulation (EC) No. 1107/2009, the EFSA Plant Protection Products and their Residues (PPR) Panel proposed that routine risk assessment for individual plant protection products should continue to use deterministic methods, and that a tiered approach to exposure assessment was still appropriate (EFSA, 2010). However it introduced an acute risk assessment for operators, workers and bystanders, for acutely toxic pesticides. The acute-acceptable-operator-exposure Level (AAOEL) reference value relates to acute non-dietary exposures that might be incurred in a single day. Guidance on more general assessments of exposure to chemicals is given in the UK by the Interdepartmental Group on Health Risks from Chemicals (IGHRC) (Anon, 2004, 2010b, 2013).

In North America, exposures to pesticides were assessed with the Pesticide Handlers Exposure Database (PHED) which provided generic mixer/loader/applicator exposure data (Krieger, 1995) until the late 1990s.

Now the Agricultural Handler Exposure Task Force (AHETF) and Outdoor Residential Exposure Task Force (ORETF) have produced pesticide handler exposure monitoring data that the Agency uses in place of PHED. Internationally work is being done to combine PHED and EUROPOEM datasets in a new North American model, the Applicator and Handlers Exposure Database (AHED) (Van Hemmen and Van der Jagt, 2005).

Many countries lack accurate information and do not have equivalent databases, so the World Health Organisation is collecting more information to improve knowledge of the extent and outcome of human exposure to pesticides, nationally, regionally and globally. WHO is developing and providing tools for the collection of internationally harmonised data on human exposures to pesticides and more efficient collection, processing and analysis of information about pesticide products. The aim is to assist countries in capacity building for prevention and management of pesticide poisoning, and in decision-making for the safe management of pesticides (http://www.who.int/ipcs/features/pesticides/en/). Ross et al. (2001) point out that studies on pesticide toxicology need to be more relevant to the occupational risk assessment.

The use of PPE does reduce dermal pesticide exposure, but compliance among the majority of occupationally exposed pesticide end users appears to be poor in many parts of the world. Yuantari et al. (2015) reporting that melon farmers in an Indonesian village were not wearing long pants and shirts with long sleeves and used the same clothes for more than 1 day without washing, and only a few protected their face with a piece of shirt material tied around their mouth. This is partly due to a lack of training and users having the less-expensive application equipment. There are also challenges of communicating information where agricultural workers have low educational attainment, low literacy rates, ethnic and cultural diversity, and language barriers in the developing world and among migrant workers in the USA (Kyriaki et al., 2014). Another major concern is that in many places, children under 18 have been allowed to help with pesticide application, a situation that needs greater enforcement of regulations to prevent it happening. International and national regulations to reduce pesticide exposure are essential, but getting small-scale farmers to adopt improved application equipment is difficult. As the majority of occupationally exposed pesticide end users are poor, their compliance in using PPE is also poor, although its use does reduce pesticide exposure. Changes in the system (i.e. changes in formulation and packaging) are easier to introduce by the agrochemical industry. More research is needed to reduce pesticide exposure and to understand the reasons for poor compliance with PPE and identify effective training methods (MacFarlane et al., 2013).

Actual exposure will vary at different stages of the treatment process, depending on the type of pesticide product being applied, for example low concentration granule or a high concentrated liquid formulation, and on the handling and application procedures adopted. Thus exposure data is usually independent of the active ingredient, but is affected by the formulation, packaging used and application equipment.

The impact of exposure will also be affected by the frequency of exposure. In Holland use of insecticides and fungicides was more frequent than herbicides as they were used 10–20 times a year on the most intensively treated crops, but single pesticide products were not used more than seven times a year (van Drooge et al., 2001). In that study, ornamental crops, such as chrysanthemums were treated more than arable crops. Some spray operators, such as those employed by contractors will be exposed for more days per year than individuals on small farms.

Potential dermal exposure is the total amount of pesticide landing on the body, including clothing, but the actual exposure to the skin will depend on the amount of deposited directly on the skin plus any that penetrates clothing and therefore available on the skin for absorption into the body. Operator exposure is significantly reduced by wearing protective clothing (Figs. 4.2 and 4.3). The basic requirement is good overalls of closely woven fabric. In temperate climates, impermeable materials are suitable and in many cases operators use disposable overalls made with a polypropylene material. Although this eliminates the need for laundering, such overalls are generally considered to be too hot to wear in tropical climates, and do not always provide as much protection as those made from cotton (Moreira et al., 1999). Various special finishes to cotton fabrics have also been tried for use in tropical countries. Gilbert and Bell (1990) have described tests for the suitability of coveralls, but laboratory tests do not always provide an accurate indication of field performance (Glass et al., 1998), while Glass and Mathers (2006) developed a test to determine the penetration of sprays through a range of materials. Simulated wear studies have been conducted to assess the impact of treating garments with a fluoroalkyl methacrylate polymer in reducing sorption and penetration of a pesticide (Shaw et al., 1996). In further field studies, Shaw et al. (2000) reported that cotton twill fabrics with and without fluorochemical finish provide barrier protection as the amount of diluted spray, which can penetrate the garment is reduced. However, without the finish, the fabric absorbs the liquid and will fail to provide protection once the fabric is saturated. This emphasises the role in engineering sprayer design to ensure it does not leak on the operator. Tests have been proposed to assess the useful life of these garments to ensure worker safety (Espanhol-Soares et al., 2013). Machado-Neto (2001) proposed that it was necessary to determine the safe work time (SWT) and exposure control need (ECN) for unsafe work conditions when the margin of safety (MOS) was less than 1. The calculation of SWT can be used to control occupational exposure when using pesticides, while the ECN permits the selection of a more appropriate safety measure depending on the working conditions. An et al. (2015) have used this procedure in valuating occupational exposure of spray applicators in glasshouses in China.

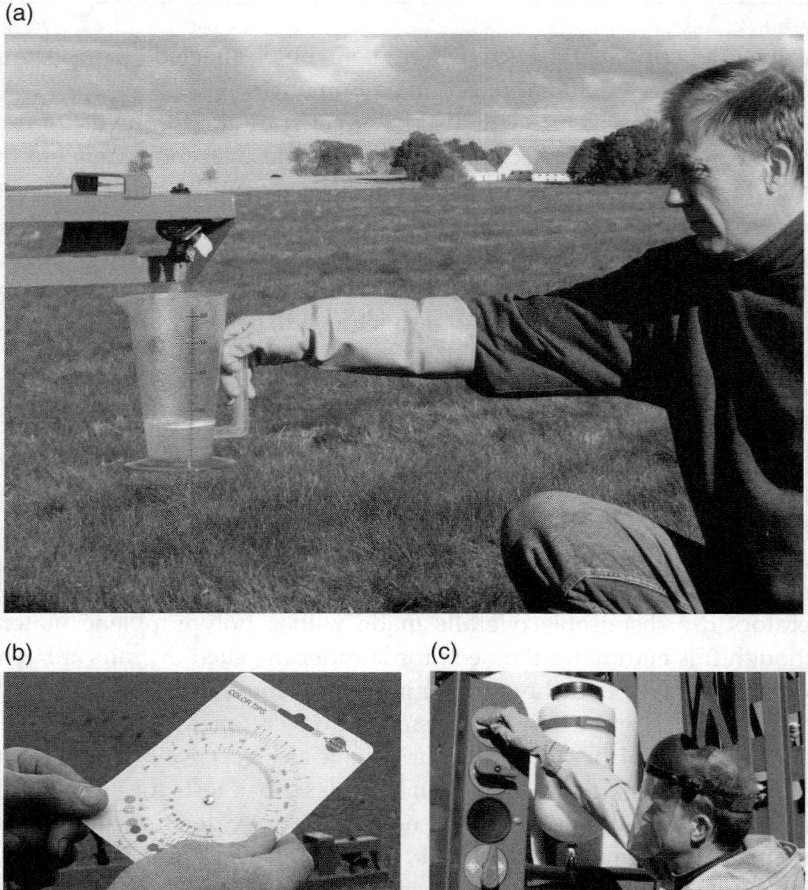

Fig. 4.2 Operating the sprayer. (a) checking nozzle output with water; (b) examining nozzle calibration chart; (c) adjusting sprayer; (d) measuring pesticide; (e) pouring pesticide into low-level hopper; (f) rinsing containers; and (g) washing hands (Hardi).

(e)

(f)

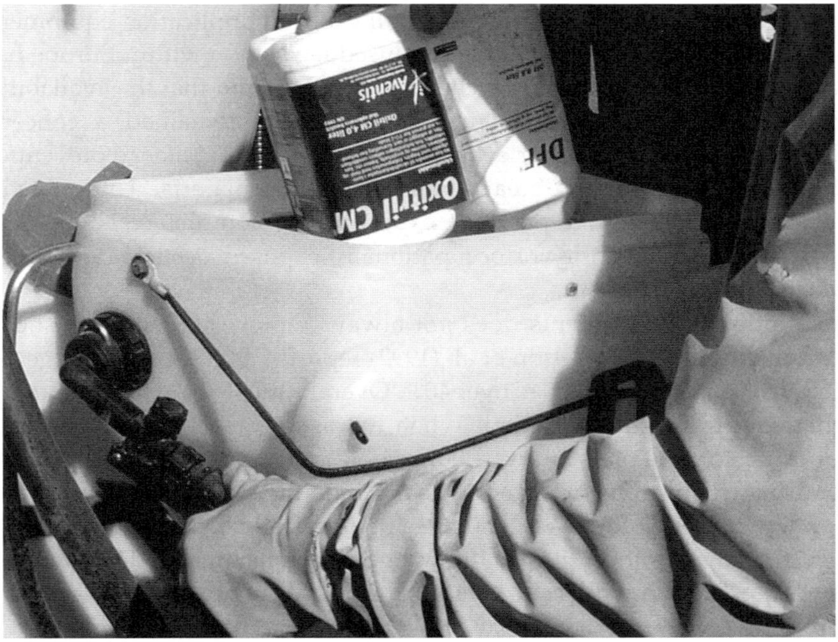

Fig. 4.2 (*Continued*)

(g)

Fig. 4.2 (*Continued*)

Spray operators find wearing protective clothing more difficult in warm climates, especially using manually carried application equipment. Glass et al. (2010) reported that harmonised legislation within Europe had been driven by the needs of the chemical industry and that the availability of PPE suitable for workers in warm climates remained a concern. They considered that further evaluation was needed to balance prevention of liquid penetration and breathability of the PPE material. Undoubtedly PPE does reduce dermal pesticide exposure, but compliance is generally poor among the majority of occupationally exposed pesticide end users (MacFarlane et al., 2013).

Laundering of garments does not always remove the entire pesticide residue in a garment. Nelson et al. (1992) reported that the percentage not removed can vary from 1% to over 40%. Overalls must always be laundered separately from other clothing. When a washing machine is used, it is important to use a pre-soak cycle if available and use the hottest water temperature with a strong detergent. The machine should then be operated for a cycle without clothes to clean traces of any pesticide from it before it is used for other clothes. Ideally the washed overalls are dried outdoors as some degradation will occur when the garments are exposed to sunlight, especially in the tropics (Shaw et al., 1997).

Wearing an apron of impermeable plastic, especially when opening pesticide containers, will protect the overalls from splashing during preparation of sprays and can be subsequently removed while spraying. Similarly

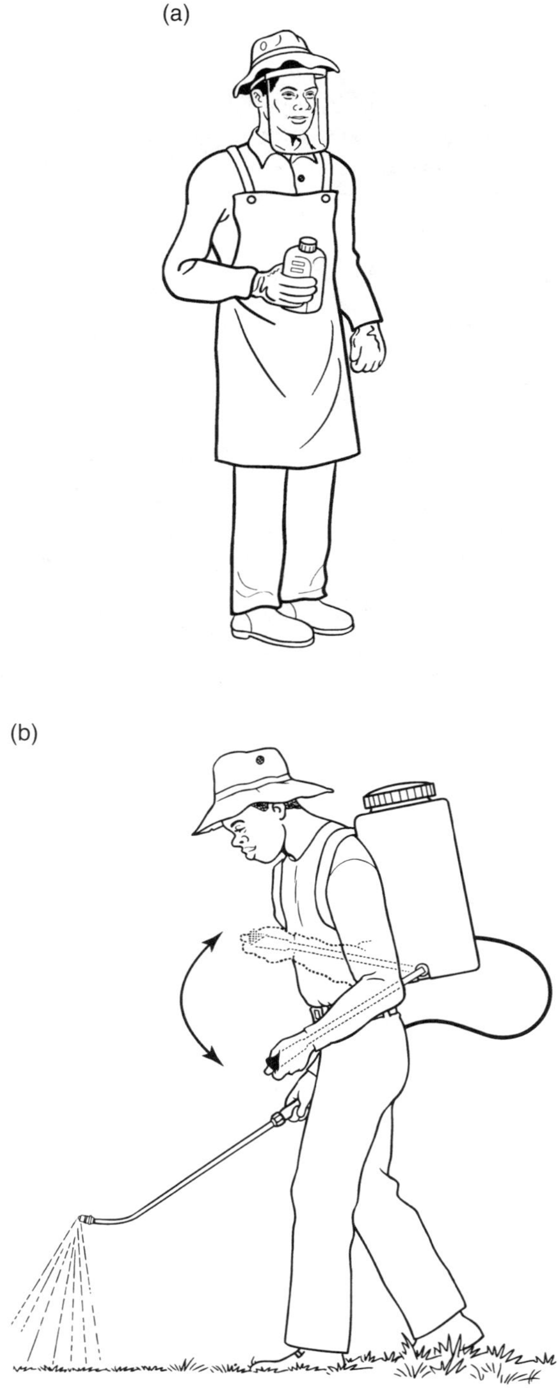

Fig. 4.3 Operator showing personal protective clothing (PPE); (a) with face mask, apron and gloves, when preparing a spray (ICI); (b) long-sleeved shirt, long trousers, hat and boots when operating a manual sprayer (NRI); (c) holding lance downwind of body to minimise operator exposure to spray (Photograph by Graham Matthews); and (d) comparison of exposure between walking into spray and holding lance downwind of the body. (*See insert for color representation of the figure*.)

(c)

(d)

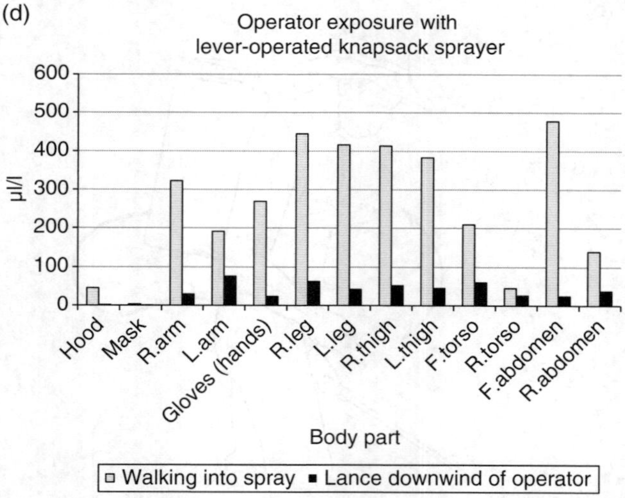

Fig. 4.3 (*Continued*)

a face shield is also recommended during mixing to protect the face and especially the eyes. Some countries prefer to recommend goggles, but these do not protect the face. When suitable overalls are not available or too expensive for small-scale farmers, the area of exposed skin should be minimised by wearing long trousers and a long-sleeved shirt. These need to be removed and washed separately from domestic laundry as soon as possible after a spray application has been completed. In Zimbabwe, many farmers fail to recognise the colour coding on containers due to lack of training, and in consequence do not protect themselves (Maumbe and Swinton, 2003).

Inevitably the hands are most likely to be exposed to sprays at all stages of application. Impermeable gloves are recommended, but generally these are uncomfortable to wear for prolonged periods. Their use is essential with the most hazardous pesticides, at least during the preparation of sprays and loading the sprayer, unless engineering controls, discussed later, provide sufficient protection. The sleeve of a coverall should cover the cuff of the glove, so that any spray on the sleeve does not run down into the inside of a glove. Poor glove hygiene and transfer to the skin or even inside of the glove when removing and replacing gloves is a big problem. Once material is inside the glove the hydrated skin conditions may lead to increased absorption and higher amounts absorbed than if gloves had not been worn. Other tests have shown that moisture increases the initial chemical transfer and skin loading providing a larger chemical reservoir for subsequent absorption (Williams et al., 2004). Where gloves are not used, there should always be a supply of water for washing any splashes off immediately. Even with gloves, the outer surface should be washed, to facilitate their removal without exposing a bare hand to pesticide. Larger sprayers are now fitted with an extra small water tank especially to allow the operator to wash gloves before removal, and to wash the hands.

Inhalation exposure is generally considered to be quite low compared with dermal exposure. This is because the amount of spray in the vicinity of the nose is generally low and the nose acts as an efficient filter so only particles in the sub 10 µm range are likely to reach the lungs (respirable fraction). Studies on devices to get pharmaceutical drugs into the lungs for asthma control have shown that larger particles are trapped in the nose (Clay and Clarke, 1987), although some of these particles in the nasal cavity are often swallowed. Very fine particles smaller than 2.5 µm diameter (known as $PM_{2.5}$s or 'respirable dust') may remain suspended in the air for weeks and once inhaled they can penetrate the microscopic air sacs (alveoli) in the lungs and cause breathing difficulties and lead to conditions such as bronchitis, asthma and emphysema. Particulate matter from combustion processes, such as exhaust from vehicles which can contribute to as much as 16% of the PM_{10} and $PM_{2.5}$ in ambient air, may be more harmful than other particles. According to WHO, total $PM_{2.5}$ mass is the best indicator for assessing air pollution-related health effects (Anon, 2010a; Buekers et al., 2014) In the UK, the Control of Substances Hazardous to Health (COSHH) Regulations on air quality set a maximum of 4 mg/m^3 (8 h time-weighted average) for respirable dust.

Many pesticides were formulated as wettable or dispersible powders, which could have more than 50% active ingredient. As the particle size is generally below 10 µm so that with a dispersant, the powder would remain in suspension when mixed with water, the operator's face could be exposed a puff of the dust. This is why these powder formulations have largely been replaced by wettable granules, or have been packaged in water-soluble plastic sachets. Another inhalation exposure concern is when pesticides are applied as fogs, with a high proportion of the droplets are considerably smaller than 25 μm, especially when applied indoors, such as in a glasshouse.

The feet should always be well protected by wearing 'rubber' boots or equivalent with the bottom of the trouser legs over the boot so that liquid or granules do not fall into the boot. In some areas spray operators fail to wear shoes; and if they do so, they are often of very often poor quality, made of absorbent materials.

The ears should be protected when the noise of the operating spraying equipment exceeds 85 decibels. This applies especially to manually carried motorised equipment fitted with a two-stroke engine, and when pulsejet fogging equipment is used.

Methodology of measuring exposure

One of the earliest methods of assessing potential dermal exposure was to attach absorbent cotton pads to different parts of the body (Durham and Wolfe, 1962). The amount of pesticide collected on 7–16 pads was determined and related to the area of relevant part of the body (Table 4.1). Hands are more frequently covered with cotton gloves. The exposure was usually reported as milligram of pesticide per hour of application, although some studies report in milligram per litre of spray applied. Absorbent pads and gloves will hold a greater volume of spray than the bare skin, but have the advantage of accumulating spray over a long period of a spray application. Placing the pads and cotton gloves under the overalls and impermeable gloves can assess the validity of protective clothing. Significant exposure can occur at the interface of a garment and skin, for example around the neck and cuffs (Anon, 2002).

Table 4.1 Area of different parts of an adult body

Body part	Area (cm^2)	% of total body area
Hands	900	4.5
Arms	2,700	13.5
Head and neck	1,200	6
Front of body	3,800	19
Back of body	3,800	19
Thighs	3,800	19
Legs and feet	3,800	19
Total	20,000	100

Pads are typically only 26 cm^2, but some are 100 cm^2 (10 cm × 10 cm), so represent a small fraction of the surface to which they are attached. Another early method was to use a strip of film placed at different parts of the body (Tunstall and Matthews, 1965). Colour dyes have been used (Fig. 4.4a) to show where the spray collected on the body. The main alternative to the pads is for the spray operator to wear a disposable overall (Chester and Ward, 1983; Chester, 1995; Sutherland et al., 1990; Machera et al., 2002; Rajan-Sithamparanadarajah et al., 2004). This can be subsequently cut into small sections, each of which is analysed separately. These studies can be with a tracer dye (often a fluorescent dye), so that the distribution of chemical in different areas can be visually assessed (Fig. 4.4b), before quantitative analysis (Lesmes-Fabian et al., 2012).

Hughes et al. (2006) used a whole body dosimetry technique, using a cotton coverall and cotton gloves as sampling media, worn over protective clothing to protect the operator showed significant difference in potential exposure while spraying captan in a small maize crop between a well-trained experienced spray operator (11.6 ml/h) and one who had not been trained (168.5 ml/h).

Special apparatus using a dodecahedron of UV lights to visualise fluorescent deposits on clothing (Roff, 1994) has been developed. Using fluorescent tracers, video imaging and pads, Fenske (1990) reported non-uniform distribution of spray deposits, which were dependent on work activity and method of application. Highest deposits were recorded on the lower part of the forearms of those preparing sprays. Incidences of high pesticide exposure may be due to pesticide spills and splashes, when mixing and loading, or during adjustments to nozzles or poor equipment maintenance arising from inadequate finance (Alavanja et al., 2001).

Blanco et al. (2005) collected field data in Nicaragua during 32 pesticide applications by observations and supplementary video-recording. A modification of Fenske's visual scoring method was used to quantify dermal exposure and showed that a wide range of factors affected the extent of exposure. Working practices explained >50% of the exposure variability, with obvious contamination of the hands, by touching surfaces such as nozzles and hoses to stop a leak, as none of the workers wore gloves.

Gao et al. (2014) compared experienced and inexperienced spray operators using a knapsack sprayer to treat maize (<80 cm high) with chlorpyrifos and reported that the total dermal exposure was 7.5× higher on the inexperienced operators. Exposure of those mixing spray was greater with emulsifiable concentrates and an oil in water emulsion compared with a wettable powder or water dispersible granules. This was principally on the hand that held the container, although in other studies it is the hand holding the container cap, being used as a measure, that is most exposed to spillage.

(a)

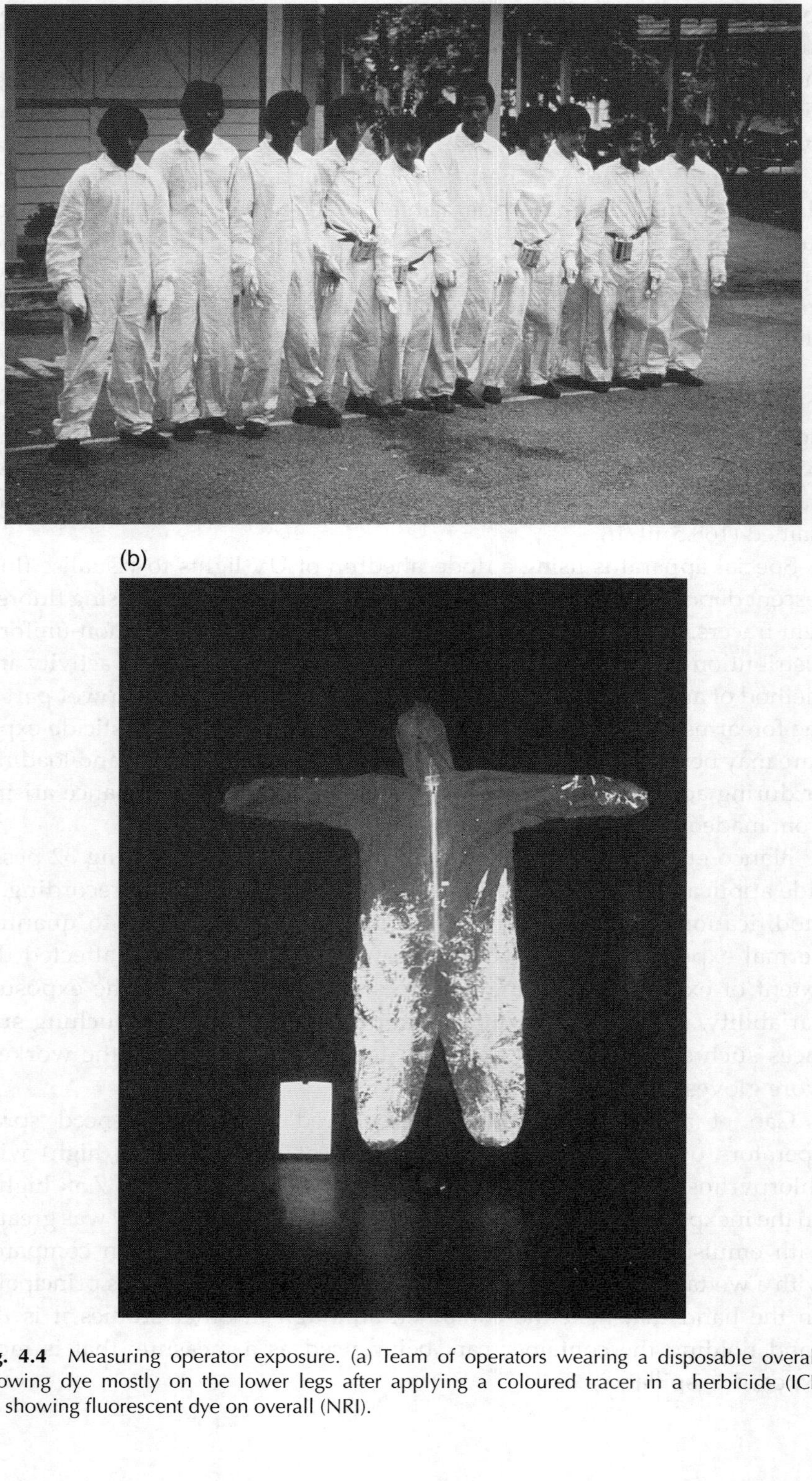

(b)

Fig. 4.4 Measuring operator exposure. (a) Team of operators wearing a disposable overall, showing dye mostly on the lower legs after applying a coloured tracer in a herbicide (ICI); (b) showing fluorescent dye on overall (NRI).

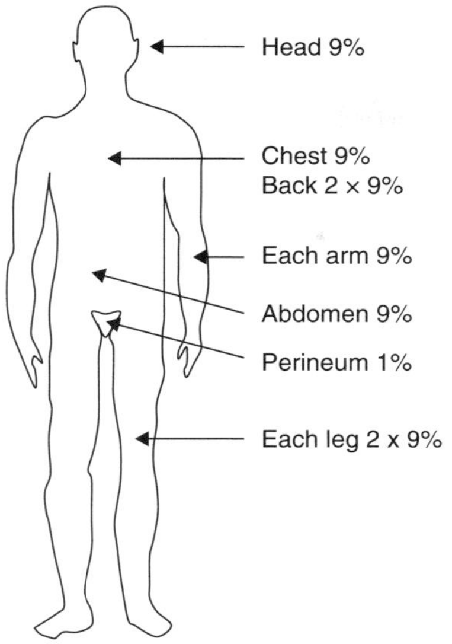

Fig. 4.4 (*Continued*) (c) The 'rule of nines'. This is primarily for a quick assessment of the area of the body affected when assessing coverage of a body by spray.

A study of dermal exposure to chlorpyrifos when treating properties in Australia to control termites indicated deposition rates of 11.1 mg/h on overalls, and 2.4 mg/h on gloves. The insecticide permeated the clothing of workers with deposition on skin patches of 0.3 $\mu g\,cm^2/h$ (Cattani et al., 2001).

In vineyard spraying, the use of a hooded or 'tunnel' sprayer reduced operator exposure compared to use of an air-assisted with a centrifugal fan (Coffman et al., 1999). A further study in vineyards sprayed using hand-held single-nozzle spray guns connected to a tractor sprayer (Tsakirakis et al., 2014) evaluated two types of protective coveralls both of which provided satisfactory protection (up to 98.4%) confirming proper use of personal protection equipment is of major importance for the operators' safety. Information on worker exposure to agrochemicals is also given by Honeycutt and Day (2001).

In assessing the surface area, the 'rule of nines' is also used (Figure 4.4c). This assumes that the head, front and rear upper and lower torso, each arm and each leg are all approximately nine per cent of the total body area. With smaller children (5 years +) the head is about 15% each arm 9.5%, each leg 17% and the front and back torso each 16%, while for a one-year-old toddler the head is again proportionally larger.

A European guidance document (EFSA, 2010; http://www.efsa.europa.eu/en/consultationsclosed/call/ppr090807.htm) has suggested that routine risk assessment for individual plant protection products should continue to use deterministic methods, and a tiered approach to exposure assessment.

Table 4.2 Example of an estimate of operator exposure (This example is to show the procedure in principle. The extent of operator exposure will vary with different equipment and the extent of operator training and care taken in practice)

	Tractor sprayer with hydraulic nozzles	Knapsack sprayer
Product concentration	250 mg/ml	250 mg/ml
Concentration in use	0.5 mg/ml	0.5 mg/ml
Spray volume	200 l/ha	225 l/ha[#]
Work rate	50 ha/day	1 ha/day
Number of tank loads	20/day	15/day
Contamination of hand	0.01 ml	3.3 ml*
per mixing operation, so with gloves	0.2 ml/day	49.5 ml/day
5% reaches the skin	0.01 ml/day	2.475 ml/day
During spraying over	5 h	8 h
Contamination of body	10 ml/h (6.5 ml on hands, 1.0 on body, 2.5 on legs)	50 ml/h (10 ml on hands, 15 on body, 25 on legs [##])
With overalls, no gloves	34.6 ml/day	116 ml/day (assumes 100% reaches the skin on unprotected hands, but 5% on body and 15% on legs[##])
Absorbed dose during mixing with gloves	0.25 mg/day (0.01 × 250 × 0.1**)	61.88 mg/day (2.475 × 250 × 0.1**)
or without gloves	5 mg/day (0.2 × 250 × 0.1**)	1237.5 mg/day (49.5 × 250 × 0.1**)
during application	1.73 mg/day (34.6 × 0.5 × 0.1**)	5.8 mg/day (116 × 0.5 × 0.1**)
Inhalation exposure	0.01 ml/h	0.01 ml/h
Inhalation absorbed dose	0.025 mg/day (0.01 × 5 × 0.5)	0.04 mg/day 100% absorption
Thus total predicted exposure	2.01 mg/day (0.25 + 1.73 + 0.025) with PPE	67.72 mg/day
which is for a person weighing 60 kg	0.033 mg/kg bw/day	1.128 mg/kg bw/day (This last value is compared with the AOEL of the pesticide being applied)

If NOAEL is 800 mg/kg bw/day, then a systemic AOEL would be set at 0.8 mg/kg bw/day. If total systemic exposure (i.e. the absorbed dose) was estimated at 0.03 mg/kg bw/day, this is 4% of the AOEL, whereas 1.128 is 141%; and not acceptable. If with the knapsack, through improved glove performance/hygiene the amount reaching the skin is lowered to 1%, this reduces the exposure to 0.31 mg/kg bw/day, that is 39% of AOEL.

*This value is based on a trial with small-scale farmers asked to measure out a dye solution with a small 50 ml plastic cup (Craig and Mbevi, 1993). Different techniques of dispensing small quantities can significantly reduce operator exposure, for example use of tablet formulation, water-soluble sachets and containers with built-in measure.

**This assumes 10% absorption through the skin.

[#]This is 15 knapsack loads.

[##]These values are used for illustration and may differ according to circumstances.

However, a requirement for an AAOEL takes account of exposure to the more toxic pesticide products during a single working day.

Examples of the types of analysis for a tractor and knapsack sprayer are shown in Table 4.2.

Exposure of hands

The hands inevitably are the part of the body most exposed to pesticides, due to handling of containers and when operating equipment. The type of exposure will depend on many factors, but in the worst case the whole hand could be coated with liquid. Substances can be removed from the surface of the skin, especially the hands by using swabs or towels, moistened with a solvent, such as 95% ethanol. The technique does not indicate what could have been absorbed before the skin was washed. Some laboratory tests were conducted by Cinalli et al. (1992) with different liquids. Using three non-aqueous liquids, subjects were asked to wipe their hands with a cloth saturated in liquid. They then used a weighed dry cloth to wipe their hands, and the amount removed was determined by the weight gained on the dry cloth. Unfortunately the dry cloth may not remove all that was on the surface of the hands, especially from between fingers or under the nails. They also determined how much was retained on hands by immersion in a container of the liquid and weighing it before and after immersion. Using a similar immersion technique with water with a surfactant Matthews (2001) reported that a bare dry hand retained on average 0.0045 ml/cm^2, whereas there was no significant increase if the hand was already wet. Excess liquid dripped off depending on the position of the hand and movement. Less water was retained on the surface of a vinyl glove. The EPA generally estimates that a man would have 6 ml retained on a hand.

To protect hands, wearing impermeable gloves is recommended, but these vary in their thickness and suitability, especially when adjusting small parts such as nozzles. Gloves with a cuff long enough to be covered by the end of the coverall sleeve are advised so that any liquid, or granule, that is on the arm does not go down inside the glove. Neoprene and nitrile gloves provide protection to a range of solvents and oils, and are suitable when using emulsifiable concentrate and similar liquid formulations. Nevertheless users should wash off any pesticide as soon as possible as some chemicals can penetrate a glove. Care is needed when washing gloves, as the rinsate subsequently acts as a source of exposure for other workers or family members, water courses, and so on. Often contamination of the gloves occurs when the operator removes the gloves using a clean hand to remove a dirty glove. Washing the gloves before removal is advised, but care is still needed when removing the glove. In some situations, especially in the tropics, an impermeable glove causes the hand to sweat. This may increase the risk of absorption. Spray operators should only apply the less hazardous pesticides and have a bucket of water readily at hand so that a bare hand, which is exposed to spray, can be washed immediately. Access to water for washing was shown to be beneficial among workers in Kenya (Ohayo-Mitoko et al., 1999).

One major cause of exposure to hands was due to metal containers with a lip around the edge. Concentrate often collected on the top of a drum after it had been tilted to pour out liquid into a measure or directly into a spray tank. Farmers, who had been dipping sheep, suffered pesticide poisoning as the drum may have used as many as eight times a day to top-up the dip tank. Exposure to a dried deposit on a surface was generally low and did not increase in proportion to the duration of contact (Glass et al., 2014).

Assessments of potential dermal exposure of horticultural and floriculture workers used a mixture of Brilliant Blue (food colouring dye) mixed with glycerine as a thickening agent to simulate a fungicide CS formulation. The same dye was used to prepare a stimulant WG formulation. Exposure of the hands was 22–62 times greater for the liquid compared to the solid formulation. Factors that influenced this were opening the aluminium seal of the pesticide container and significant external surface contamination of pesticide containers (Berenstein et al., 2014). The latter is often due to a residue left in the container cap, used on small farms and amateur gardeners to measure the amount of pesticide needed for a single sprayer load (Craig and Mbevi, 1993); Harrington et al., 2005). This method of measuring inevitably results in the operator being exposed to the concentrate as few wear gloves. Using the cap is also a source of errors in concentration which varied from 55 to 177% of the intended concentration, in one series of tests. This fault can be overcome by supplying the correct dose for one sprayer load in a sachet. In the home garden situation, exposure of the operator to diluted spray was typically 20 ml/h, with up to 10 mg/h of the active substance, while spraying for 5–10 min (Harrington et al., 2005). In order to reduce the need for personal protective clothing, emphasis has been placed on engineering controls to minimise exposure. The trend for small gardens has been therefore to market ready-to-use products.

Inhalation exposure

In the open air, the risk of inhaling spray droplets is extremely low. Most sprays contain only a small fraction of the volume in droplets smaller than 100 µm. While these small droplets can shrink, especially on hot days and with low humidity, the smallest droplets in the 1–10 µm range that could be inhaled are readily carried downwind and away from the spray operator. Any larger droplets close to the nose may be deposited on the face or filtered within the nose and would not reach the lungs. The situation is different when applying pesticides inside buildings, stores and glasshouses where small droplets can remain airborne close to the operator.

Berger-Preiß et al. (2005) studied inhalational and dermal exposures in different settings to generate exposure data for health risk management. From this data a model was developed that calculates the airborne concentration of the active substance based on the concentration of the active substance in the preparation, droplet dispersion due to turbulence, droplet settling and evaporation kinetic of the solvent.

Protection from small airborne droplets (<25 µm diameter) is necessary when fogging. A respirator is then essential with suitable filters to remove the small droplets. Care is needed that the filters are still effective and, if there is any doubt, new filters suitable for the pesticide being used should be fitted. Disposable face-masks are not respirators and are only useful if dusts are applied with larger particles or the operator wishes to reduce dermal exposure in the vicinity of the mouth and nose. Specialised respirators and masks cover the whole head and have a forced air system that pumps filtered air over the operator's face.

Inhalation exposure studies

The exposure to airborne spray is ideally carried out using personal monitors, which have a pump system to draw air through a filter (Wolfe, 1976). The filter may be a simple cotton gauze or adsorbent resin. In both cases it is important to know the breathing rate of the person and the volume of air sampled to interpret the potential inhalation of a pesticide. Personal monitors mounted in the breathing zone of an operator are calibrated to have a flow of 1–2 l of air per minute. Bjugstad and Torgrimsen (1996) using a respiration rate of 1.75 m^3/h calculated exposure when using several types of application equipment in a greenhouse. The highest respiratory exposure was, as expected, when using a thermal fogger, hence the need for using respiratory protective equipment (RPE). In pest control operations by professionals, airborne concentrations of permethrin were highest when dusts were applied, especially when treatment was above the operators, and when several worked together in a confined area (Llewellyn et al., 1996).

Currently when using spray drift data, it has been suggested that different percentiles of spray drift data should be used to reflect the different time periods of exposure, namely the 95th percentile for acute assessments and the mean for longer term assessments.

In the EFSA guidance for outdoor spraying downwards, an example of the milligram exposure per kilogram-active substance applied is suggested in Table 4.3 for operators during application. Other examples are given in the EFSA document (EFSA, 2010, 2014).

Table 4.3 Contrast between exposure for a person using a tractor and a knapsack sprayer

	Dermal exposure		Inhalation exposure	
	75th centile	95th centile	75th centile	95th centile
Tractor boom sprayer	Hands 0.730 Body 0.917	Hands 10.6 Body 4.71	0.0107	0.0781
Knapsack sprayer	Hands 611 Body 1777	Hands 2856 Body 10949	0.783	5.99

Biomonitoring

Assessments of exposure determine how much reaches a person and depending on the extent to which the body is covered by clothing, how much is actually on the skin. Human skin protects the body very effectively from chemicals, especially in relation to the skin of other animals such as rats, but a proportion will be absorbed through the skin and reach the blood stream. The pesticide like other chemicals in the body will reach the liver and be subject to breakdown and metabolites are excreted via the kidney. To determine how much has entered the body, it is usual to analyse urine samples. Analytical techniques such as chromatography and mass spectrometry are used to measure accurately urinary metabolites or blood body burdens of several classes of pesticides (Barr, 2008), due to occupational exposures to pesticides, as well as general background exposures from residential or dietary exposures. Biomonitoring is important not only to prove that workers and consumers are not exposed but also to show that chemical compounds are not accumulating in the body or seriously affecting any body functions. Nevertheless the techniques are used as routine procedures in only a few countries (Tsatsakis et al., 2012). Urinary analyses of children showed that frequent consumption of apples, fruit juices and chicken/turkey meats gave significant levels of TCP, indicating chlorpyrifos exposure and 3-PBA from permethrin exposure, two insecticides frequently used to control insect pests on a variety of agricultural crops (Morgan and Jones, 2013). However, urinary metabolic concentrations were reduced to non-detected or close to this when 'organic' fruit were consumed over a 5-day organic diet intervention (Lu et al., 2008).

Apart from urinary analyses, the presence of OP metabolites has been examined in hair samples. Tsatsakis et al. (2010) developed a simple, fast, precise and highly sensitive (LOQ lower than 20 pg/mg) method for the simultaneously quantitation of four non-specific metabolites of OPs pesticides in hair. Testing hair can be used to assess chronic OP exposure by the detection of OP metabolites (Margariti and Tsatsakis, 2009). The possibility of using saliva samples to monitor exposure to some pesticides has also been considered (Fenske and Day, 2005; Nigg et al., 1993). Denovan et al. (2000) considered that collecting samples of saliva was an easy field method and was sensitive and inexpensive way of assessing atrazine exposure. Wang et al. (2008) reported a new carbon nanotube-based electrochemical sensor to detect cholinesterase salivary activity and thus provide a sensitive and quantitative tool for non-invasive biomonitoring of exposure to organophosphate pesticides. However saliva sampling is not suitable for all pesticide biomonitoring, especially when rapid metabolism occurs. Sampling blood serum indicated that evaluated workers, treating tomatoes and ornamental ferns in Florida, USA, are effectively protected by the PPE they were wearing against exposures known to result in acute toxicity (Johnson et al., 2014).

Some biomonitoring studies have been done with a low-toxicity pesticide and its metabolites measured in urine samples (e.g. Krieger and Dinoff, 2000). Skin moisture will affect absorption, as indicated by increased dermal absorption of propoxur under conditions of high humidity and 30°C temperature (Meuling et al., 1997). Investigating the relation between exposure doses in the field and cumulative excretion of biomarkers in urine, Aprea (2012) confirmed that respiratory exposure of agricultural workers to pesticides during application and re-entry of treated fields or greenhouses is a much smaller percentage than that due to skin exposure, which contributes more than 90% of the real dose.

Urine samples from sheep dippers pre- and post-dipping with an OP insecticide showed a high level of urinary metabolites when compared with office workers and others (Table 4.4) with no known occupational exposure. The metabolites detected in office workers may have been due to domestic use or from dietary sources (Nutley et al., 1995). Total pesticide OP metabolites in samples from a person, who made an unsuccessful suicide attempt, were 1000× greater than those from sheep dippers. Most exposure was due to the concentrate, thus 100 µl of concentrate would be equivalent to exposure to 150 ml of the diluted pesticide in the dip. Unfortunately, very few of those dipping sheep wore the recommended protective clothing (Buchanan et al., 2001). Inhalation was a minor route of exposure.

Studies with diazinon showed that 40% of the OP was excreted in the first 24 h, but then there is a slow rate of about 5% per day between days 3 and 7 (Anon, 1999b; Wester et al., 1993). In Spain, urine samples from operators exposed to the insecticide acetamiprid applied with a spray gun in greenhouses also showed that excretion was primarily within 24 h (Marin Juan et al., 2004) (Fig. 4.5). Similarly in tests with the OP propetamphos, a urinary metabolite was shown to be a suitable bio-marker, with excretion

Table 4.4 Urinary dialkyl phosphate data from various occupational groups and workers (Adapted from Nutley et al., 1995)

Occupation	Mean (range) total urinary 'ethyl' phosphate metabolites/ mmol/mmol creatine	90% results less than	Number of samples/ individuals
Office workers	1.6 (0–32)	6	106/106
Sheep dippers			
Pre-dipping	5.6 (9–162)	14	159/159
Post-dipping	15 (0–189)	40	337/167
Agricultural workers			
Pre-exposure	1.3 (0–17)	5	35/35
Post-exposure	10.3 (0–159)	27	59/35
Formulators	42.8 (0–479)	72	88/10

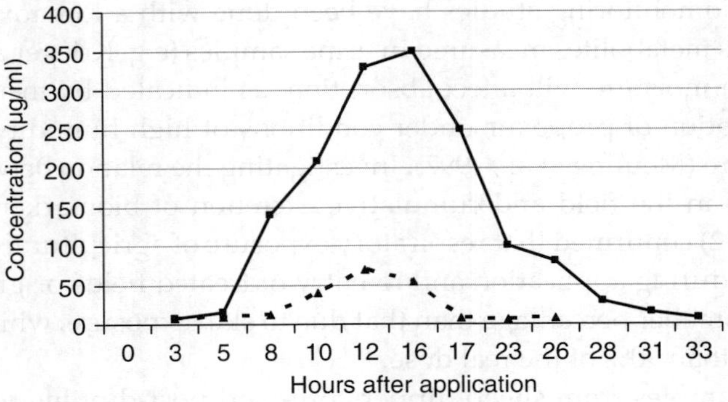

Fig. 4.5 Biomonitoring urine samples showing effect of wearing PPE. Solid line without PPE; dotted line with PPE (Adapted from Marin Juan et al., 2004).

over a longer period following dermal compared to oral exposure (Garfitt et al., 2002). No adverse health effects were to be expected following biomonitoring of acetpchlor, although tractor drivers using open cabs were more exposed than those who had a closed cab (Gustin et al., 2005).

Biomonitoring has been used in assessing exposure of persons using small application devices inside residences, where exposures are considerably less and occur over a longer period than in an open-field agricultural use (Kreiger et al., 2001). Generally measurements indicate exposure more accurately as they reflect the extent of dermal absorption through the skin over the whole body and the effect of excretion of the pesticide. In addition to urinary monitoring, the possibility of using saliva samples to monitor exposure has been considered (Donovan et al., 2000; Fenske and Day, 2005).

The Committee on Toxicology (COT) Working Group on Organophosphates considered whether single, prolonged or repeated exposure to low doses of organophosphate compounds (OPs) can cause long-term adverse health effects (Anon, 1999c). Other sources of information on assessing pesticide exposure include Chester (1993), Curry et al. (1995) and the International Centre for Pesticides and Health Risk Prevention, a Collaborating Centre on Occupational Health, specialising in pesticides, which is based in Milan, and publishes a newsletter *Pesticide Safety News* in English, Italian and Spanish.

First aid

If a spray operator becomes ill, while working, the doctor must be informed of the name of the active ingredient and given as much information as possible by showing the doctor a leaflet or label about the chemical.

Treatment by a doctor will depend very much on the type of poisoning. When using an organophosphate or carbamate (anticholinesterase), an injection of atropine is useful, but suitable antidotes for organochlorine poisoning are not available. A person, who has ingested liquid that contained paraquat, can be treated by ingesting large quantities of Fuller's earth, which absorbs the herbicide. Morphine should not be given to patients affected by pesticide poisoning. A first-aid kit and a supply of clean water for drinking and washing any contaminated areas of the body should be readily available. Some of the symptoms of poisoning such as a headache may be due to effects of working for a long period in direct sunlight without sufficient protection, such as wearing a hat, and to dehydration by not drinking sufficient water. On large-scale spraying programmes first-aid kits should be carried in vehicles and aircraft. People regularly involved in applying organophosphate pesticides should have a routine medical examination to check the cholinesterase levels in their blood plasma.

A cholinesterase monitoring rule was introduced in Washington State, USA, in 2004 as there was concern about the effect of organophospate and *N*-methyl carbamate sprays being applied. During 2004, following baseline tests carried out on 2630 employees, 20% of the 580, who were monitored, experienced cholinesterase depression, necessitating the employer to evaluate pesticide handling practices. By 2010, the number with cholinesterase depression has been reduced to 3%, indicating that the programme had enabled the state authorities to monitor worker safety and health better. The monitoring was not linked to any specific pesticide, application method or crop being treated, but some of the poisonings occurred following the use of tractor-mounted air-blast orchard sprayers. Irrespective of the cause of poisoning, it emphasised the importance of engineering controls and working practices, including correct protective clothing to minimise exposure. A change to using less-toxic pesticides is indicated whenever possible as this would reduce the need for routine monitoring.

In many countries there are dedicated poisons centres from which medical doctors can get advice on treatment. Information about poison centres in the USA are given by Calvert et al. (2010). There are also web pages giving advice such as http://www.ccohs.ca/oshanswers/chemicals/pesticides/firstaid.html and http://psep.cce.cornell.edu/Tutorials/core-tutorial/module10/index.aspx

In contrast some tropical countries lack the expertise (Ngowi et al., 2001a), yet it is in these countries that some of the most toxic pesticides are still used without trained staff. Thus occupational poisoning has continued to be a serious problem among farm workers, for example, those in the coffee growing areas of Tanzania (Ngowi et al., 2001b).

Periods of exposure

Spray preparation

Most pesticides are applied as sprays so the main concern is when the concentrate is diluted to prepare a spray. There used to many problems when opening containers and pouring pesticide due to 'glugging' and splashes, but significant changes in container design have improved the ease of pouring pesticide. Containers larger than 1 l usually have a standard thread opening of external 64 mm diameter. Widespread adoption of low-level induction bowls has also facilitated the transfer of the formulation into a sprayer. Glass et al. (2002a) reported that usually less than 0.01 ml per fill of the sprayer was on the operator, when pouring pesticide from a container into an induction hopper. The use of an apron and face shield at this stage can avoid splashes on the coveralls or face. There is little risk of creating small droplets which could be inhaled during mixing and loading so inhalation exposure is minimal (Wolf et al., 1999). A tray under the induction bowl when spray is being prepared is used to collect any spillage on the ground.

Concern about operators climbing up on to a sprayer tank to pour concentrate from drums into the tank opening has led to the design and adoption of low level induction bowls that must be no higher than 1 m above the ground. The induction bowl makes it easier for operators to add the spray concentrate and as they are fitted with rinsing nozzles, the empty containers can be thoroughly washed (triple rinse) before disposal. Studies confirmed that spillages were reduced by using the induction bowl, which is now used on 82% of the arable area of the UK (Garthwaite, 2002).

Further reduction of operator exposure is possible with adoption of closed transfer systems (Glass et al., 2002b). Initial tests with a liquid formulation have shown a significant decrease in the amount of liquid that is collected on the hands and overalls (Table 4.5) (Glass et al., 2004). An evaluation in glasshouses using manual sprayers showed that the mix/load operation was the stage with highest exposure (Ramos et al., 2010).

Table 4.5 Volume of liquid (ml) from field studies when using induction bowl or a closed transfer system

	Mean value*	Maximum value
Induction bowl		
Body	0.07	0.65
Hands	0.25	1.05
Closed transfer system		
Body	0.01	0.03
Hands	0.03	0.06

* Note minimum values were below the level of detection.

Ideally the pesticide is transferred to the sprayer by a closed transfer system, incorporating a measure that meets the BS 6356 part 9. So far these closed transfer systems have been used for specialist or certain high-volume uses (Fig. 4.6). Many pesticides are not available in suitable containers, with

(a) (b)

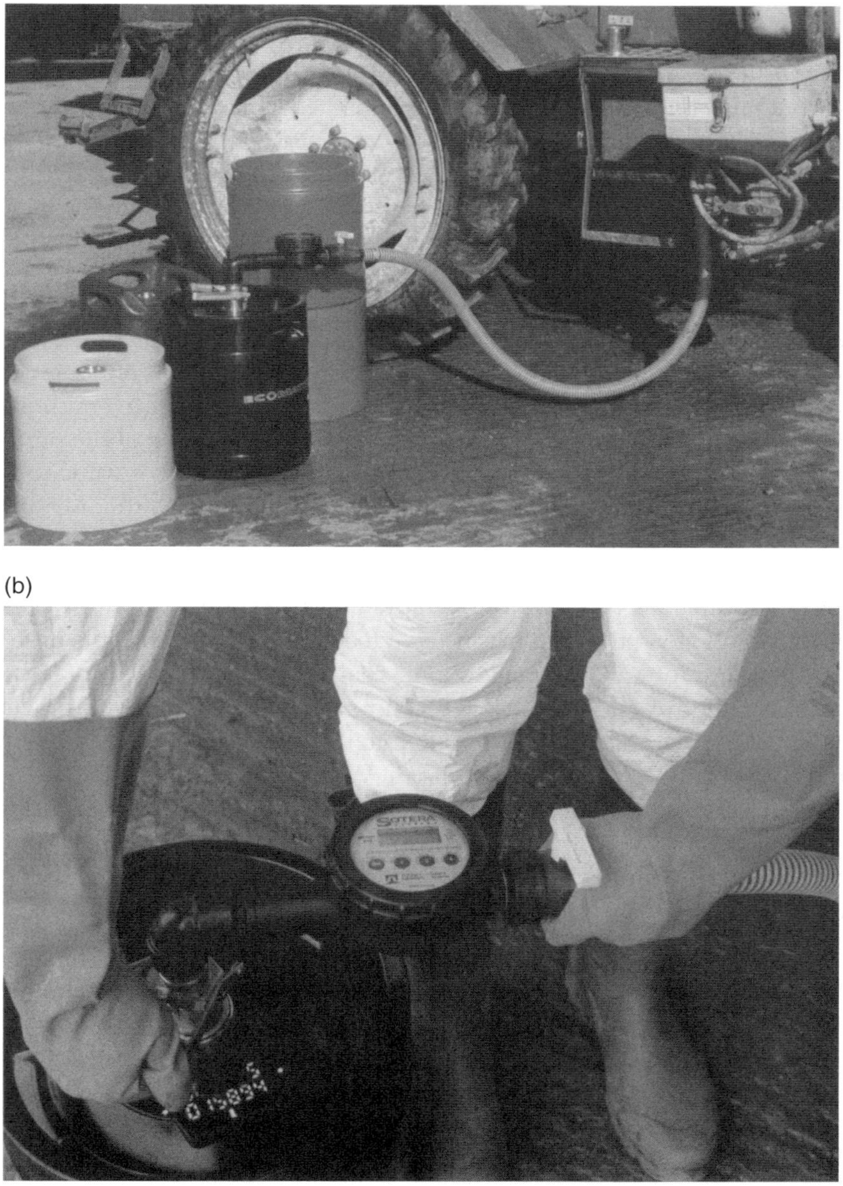

Fig. 4.6 Closed transfer system. (a) Showing drum connected to sprayer. (b) Close up of drum with operator attaching it to the sprayer (WISDOM).

the chemical industry appearing reluctant to support multi-trip containers in many countries due to cost implications. Provision of different formulations, including wettable granules as well as liquids, makes it difficult for a standard system. However, some closed transfer systems do fit standard containers, and allow the empty container to be washed out.

During spraying

In large-scale agriculture, the operator is well protected inside the tractor cab, but must be careful not to take contaminated clothing into the cab, supplied with filtered air. In well-designed cabs, filters remove at least 99% of aerosol particles larger than 3 μm (Hall et al., 2002). Care is needed if nozzles, the spray boom or other components need attention, when the hands in particular could be exposed to concentrated spray deposits. Touching surfaces in the cab then transfers pesticide to the steering wheel and seats (Landers, 2004). Most sprayers now have a separate water tank for washing any spray off gloves and separate lockers for clean and used PPE. Actual exposure during this period should therefore be minimal, although the period during which the operator may be exposed can be several hours. Even when approved coveralls are used, the correct wearing of the protective clothing is an important factor determining its protective value to the wearer, as demonstrated by research sponsored by the Health and Safety Executive in the UK (HSE, 2002).

In many parts of southern Europe and elsewhere and in glasshouses, the traditional method of high volume spraying has continued. In Holland, typically a motorised pump was used to feed spray along a flexible hose at high pressure (often 20–30 bar) to a manually directed 'pistol'. Spray volumes usually exceed 1000 l/ha. A second person may assist with moving the long hose. The operator may walk into the area treated and can be covered with spray. Data (Table 4.6) from de Vreede et al. (1998), with a period of about 9 min for mixing and loading followed by an average of 81 min for application, confirmed that the hands were the most exposed part of the body, especially during mixing and loading the sprayer. Of all their samples (190) mean penetration through the overalls was on average less than 5%.

Table 4.6 Mean amounts of active substance (mg/h detected on spray operators using high volume application (geometric standard deviation)

	Mean	GSD
Inhalation	5.1	(5.0)
Gloves (mixing/loading)	13,110	(7.1)
Gloves (application)	760	(4.9)
Overall	1,710	(3.1)
underwear	40	(4.4)
Socks	7.5	(3.3)

Walking backwards away from the spray reduced exposure by a factor of 7×. Operator exposure was also considerably reduced when a small vehicle or trolley mounted sprayer was used instead of the long hose (Nuyttens et al., 2004, 2009) (Figs. 3.2f and 4.7).

On small farms with manually carried equipment, the situation is quite different. This is especially true in tropical countries, where relatively few farmers can afford to use PPE and find it uncomfortable in a hot climate, even when applying highly toxic pesticides. Many operators have had only primary education or are illiterate, so have little knowledge about the pesticides they are applying (Mekonnen and Agonafir, 2002). Pesticides may be stored in their houses and may not wear PPE unless it is supplied or advised by a local farming association (Al Zadjali et al., 2015). Operators are not only exposed to the diluted spray for lengthy periods as they tend to walk through crops with the spray nozzle held in front of them, but poor-quality equipment and lack of maintenance often results in leaks of pesticide over the operator's hands and back. FAO has issued minimum requirements for sprayers in an effort to get better quality sprayers used (see http://www.fao.org/docrep/X2244E/X2244E00.htm).

Early studies indicated that placing the nozzle behind the operator would significantly reduce exposure to pesticides (Fernando, 1956; Tunstall and Matthews, 1965), but few farmers have accepted this. The exposure of an operator, walking through cotton with both a knapsack and with the lance held in front of their body – as most operators do – is on the lower legs.

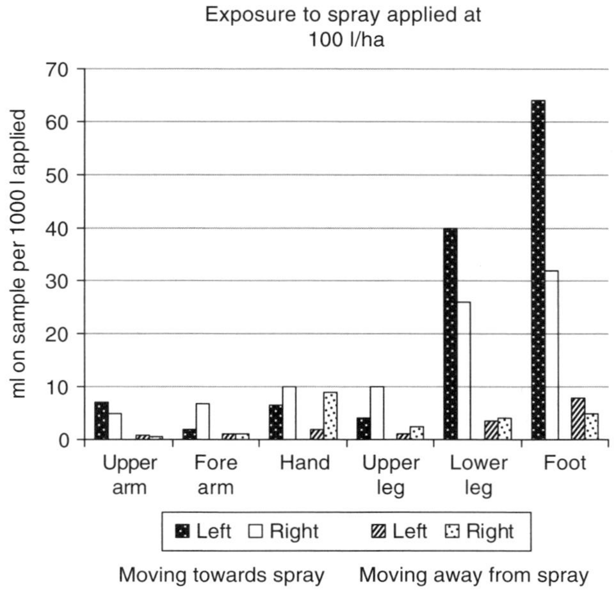

Fig. 4.7 Contrast when moving towards and away from the spray in a glasshouse (Adapted from Nuyttens et al., 2004. © Association of Applied Biologists).

In Argentina, operators using a knapsack sprayer, were exposed to 78 ml/h and 12 ml/h for treating Swiss chard and lettuce crops, respectively, with most on the lower legs and gloves (Hughes et al., 2004). This exposure can be reduced by holding the lance downwind of the operator as recommended with the spinning disc sprayer (Table 4.7) (Thornhill et al., 1996). The spinning disc sprayer applies a higher concentration of spray, but the volume applied is far lower, and the person holding the sprayer does not walk through treated foliage. Thus the next exposure shown in Table 4.7c is similar to that indicated in Table 4.7a with the lance on a knapsack sprayer held downwind of the operator.

Operators of manually carried equipment are often exposed to spillages and leaks from trigger valves, especially if the sprayers are poorly maintained. Farmers often complain that spare parts are difficult to obtain so they contrive to repair sprayers with inappropriate materials. There is a raised flange on the side of the tank closest to the straps on some sprayers to protect the operator from spillages from the sprayer. In some countries, a simple plastic tabard has been supplied to reduce direct exposure of the body yet provide adequate ventilation. Protection of the legs and feet was required for users of knapsack sprayers applying the herbicide paraquat, as most potential dermal exposure was on these areas of the body (Machado-Neto et al., 1998). A study in Malaysia confirmed the importance of protective clothing to reduce exposure (Baharuddin et al., 2011).

Ohayo-Mitoko et al. (1999) reported in Kenya that acetylcholinesterase inhibition was greater in those spraying compared to mixers, presumably because they were exposed over longer periods. The study also confirmed workers using WHO Class III pesticides were less affected than those spraying more toxic pesticides.

Table 4.7 Potential operator exposure with a lever-operated knapsack sprayer in cotton (a) holding the lance downwind and (b) in front of the operator, and (c) a spinning disc VLV sprayer (ULVA+) with disc downwind of the operator, expressed as μl/l of spray applied as determined from fluorescent tracer study (Data from Thornhill et al., 1996)

Part of body	Area of disposable overall (cm^2)	Deposit (a) (μl/l)	Deposit (b) (μl/l)	Deposit (c) (μl/l)
Hood (head)	1,200	1.8	45.6	9.3
Mask	172	0.7	3.2	0.05
Right arm	1,350	29.7	322.5	63.1
Left arm	1,350	76.3	191.0	133.0
Gloves (hand)	900	23.6	269.4	33.6
Right leg	1,250	62.7	444.3	11.9
Left leg	1,250	42.6	416.2	21.3
Right thigh	1,900	52.6	413.3	13.1
Left thigh	1,900	45.9	383.2	6.1
Front torso	2,750	60.9	209.3	33.9
Rear torso	2,750	26.2	45.7	30.4
Front abdomen	3,550	25.0	477.4	39.7
Rear abdomen	3,550	38.0	139.7	65.8
Total	23,872	486.0	3360.8	461.25

Exposure to those who only wear a long-sleeved shirt and long trousers or the equivalent local clothing can be minimised by controlling spray pressure. Operator exposure was reduced by using a control flow valve that controlled the pressure to avoid both too fine or too coarse a spray (Shaw et al., 2000). Applying a coarser spray from an air induction nozzle can also reduce operator exposure (Stephens et al., 2006; Wicke et al., 1999).

In semi-arid areas, ULV spraying has been extensively adopted on cotton, although more recently to follow IPM, there was been a change to VLV, generally using water-based formulations in about 10 l/ha, as this allowed more choice of insecticide. When highly toxic ULV formulations applied, a study by Kummer and van Sittert (1986) confirmed by biomonitoring, indicated an increased absorption of insecticides, though no clinical signs or symptoms of intoxication were observed. Most exposure was during filling the sprayer from containers.

Potential operator exposure was higher when spraying strawberries in a tunnel system than in an 'open-field' (Bjudstad and Hermansen, 2010).

After spraying

Care is also needed after spraying. Any of the pesticide product unused needs to be returned to the store, and all equipment must be washed out after use, otherwise residues will accumulate and quantities subsequently washed off may be harmful if they enter a watercourse (Ramwell et al., 2004) The boom had more pesticide than the external surface of the tank (Ramwell et al., 2006). By careful calibration, all the dilute spray will have been used, so washing of the tank can be done within the treated field. By having extra water available in the field the washings of the sprayer tank can be applied by spraying on the last swath. However, low-pressure systems on sprayers used for infield cleaning perform inadequately compared with high-pressure systems used in a farmyard which typically can remove 95% of dried chemical deposits on the external surfaces of field crop sprayers (Cooper et al., 2006).

Efforts are being made to get manufacturers of sprayers to examine the design of sprayers so that the amount of liquid that remains in the sprayer, pump and associated hoses is kept to a minimum. However, parts of the sprayer, including inaccessible areas of the pump, will have pesticide deposits, so those doing the maintenance of equipment need to be careful when dismantling parts of sprayers (Fig. 4.7).

In some countries where diluted pesticide contaminated water has to be disposed of, special areas known as 'biobeds' are used (Fogg and Carter, 1998; Basford et al., 2004) (Fig. 4.8). One type of biobed consists of a lined pit filled with a mixture of 50% straw, 25% peat and 25% soil and covered with turf grass. Fogg et al. (2003a) reported that biobeds offered viable means of

Fig. 4.8 Washing the sprayer on a bunded area draining into a biobed in the foreground (Photograph by W. Basford).

degrading some mixtures of herbicides and insecticides, although it may be necessary to avoid releasing certain pesticides to a biobed. Water management is crucial with lined beds to ensure sufficient moisture for microbial degradation, yet avoid overflow from the bed. With unlined beds, the most mobile pesticides were liable to leaching, although >99% was removed by the system and degraded within 9 months (Fogg et al., 2003b). The siting and type of biobed is important, so advice is needed to comply with the waste regulations. Pesticide detected in drainage from a biobed was 0.1% of that caused by run-off from a concrete surface. Farm-scale effluent treatment plants can also be used to remove small quantities of pesticides from water (Harris et al., 1991).

Traditionally, farmers disposed most of their waste in on-farm dumps and by burning packaging where possible, but the Waste Management Regulations 2006 in the UK now set out the methods for disposal of pesticide waste and packaging. Advice in the UK is available at http://www.pesticides.gov.uk/Resources/CRD/Migrated-Resources/Documents/C/Code_of_Practice_for_using_Plant_Protection_Products_-_Section_5.pdf

Triple-rinsed containers should be disposed of through a licensed waste-disposal contractor. A global perspective of recycling empty pesticide containers (Jones, 2014) pointed out the need for countries to have appropriate regulations that set out methods of collection and the location of collection sites. Stewardship schemes (Denny, 2013) have made progress in some countries, such as Brazil, where a very high proportion of plastic containers are now recycled (Rando, 2013). Careful management of individual pesticide stores is crucial to minimise the problem of obsolete stocks (Dollimore and Schimpf, 2013) and waste disposal. Guidance on obsolete stocks is at http://www.fao.org/agriculture/crops/obsolete-pesticides/prevention-and-disposal-of-obsolete-pesticides/en/

This chapter has shown that with the correct choice of pesticide, packaging and equipment, together with use of appropriate protective clothing, operators can apply pesticides safely with minimum exposure to the chemical used.

References

Al Zadjali, S., Morse, S., Chenoweth, J. and Deadman, M. (2015) Personal safety issues related to the use of pesticides in agricultural production in the Al-Batinah region of Northern Oman. *Science of the Total Environment* **502**, 457–461.

Alavanja, M.C.R., Sprince, N.L., Oliver, E., Whitten, P., Lynch, C.F., Gillette, P.P., Logsden-Sacket, N. and Zwerling C. (2001) Nested case–control analysis of high pesticide exposure events from the agricultural health study. *American Journal of Industrial Medicine* **39**, 557–563.

An, X., Ji, X., Jiang, J., Wang, Y. and Wu, C. (2015) Potential dermal exposure and risk assessment for applicators of chlorthalonil and chlorpyrifos in cucumber greenhouses in China. *Human and Ecological Risk Assessment* **21**, 972–985.

Anon (1999a) Dermal exposure to non-agricultural pesticides: exposure assessment document Health and Safety Executive, London.

Anon (1999b) Risk assessment to risk management: dealing with uncertainty. Risk assessment and Toxicology Steering Committee, Institute for Environment and Health, Leicester.

Anon (1999c) Organophosphates. Committee on Toxicity of Chemicals in Food, Consumer Products and the Environment, Department of Health, London.

Anon (2002) Dermal exposure resulting from liquid contamination. HSE/DSTL Report A-2248. Health and Safety Executive, London. Available at: www.hse.gov.uk/research/rrpdf/rr004.pdf (accessed on 6 May 2015).

Anon (2004) Guidelines for good exposure assessment practice for human health effects of chemicals. IGHRC Cr 10. Medical Research Council Institute of Environmental Health, University of Leicester, Leicester.

Anon (2010a) *Traffic-Related Air Pollution: A Critical Review of the Literature on Emissions, Exposure, and Health Effects*. Special Report 17. Health Effects Institute, Boston, MA.

Anon (2010b) Current approaches to exposure modelling in UK Government Departments and Agencies. Institute of Environment and Health, Cranfield University, Cranfield.

Anon (2013) Predictive approaches to chemical hazard identification and characterisation: current use by UK Government Departments and Agencies. Institute of Environment and Health, Cranfield University, Cranfield.

Aprea, M.C. (2012) Environmental and biological monitoring in the estimation of absorbed doses of pesticides. *Toxicology Letters* **210**, 110–118.

Baharuddin, M.F.B., Sahid, I.B., Mohd, M.A.B., Sulaiman, N.N. and Othman, F. (2011) Pesticide risk assessment: a study on inhalation and dermal exposure to 2,4-D and paraquat among Malaysian paddy farmers. *Journal of Environmental Science and Health B* **46**, 600–607.

Barr, D.B. (2008) Biomonitoring of exposure to pesticides *Journal of Chemical Health & Safety* **15**, 20–29.

Basford, W.D., Rose, S.C. and Carter, A.D. (2004) On-farm bioremediation (biobed) systems to limit point source pesticide pollution from sprayer mixing and wash-down areas. *Aspects of Applied Biology* **71**, 27–34.

Berenstein, G.A., Hughes, E.A., March, H., Rojic, G., Zalts, A. and Montserrat, J.M. (2014) Pesticide potential dermal exposure during the manipulation of concentrated mixtures at small horticultural and floricultural production units in Argentina: the formulation effect. *Science of the Total Environment* **472**, 509–516.

Berger-Preiß, E. Boehnckey, A., Konnecker, G., Mangelsdorf, I., Holthenrich, D., Koch, W. (2005) Inhalational and dermal exposures during spray application of biocides. *International Journal of Hygiene and Environmental Health* **208**, 357–372.

Bjugstad N. and Hermansen, P. (2010) Potential operator exposure when spraying in a strawberry tunnel system. *Aspects of Applied Biology* **99**, 139–147.

Bjugstad N and Torgrimsen, T. (1996) Operator safety and plant deposits when using pesticides in greenhouses. *Journal of Agricultural Engineering Research* **65**, 205–212.

Blanco, L.E., Aragaón, A., Lundberg, I., Lidén, C., Wesseling, C. and Nise, G. (2005) Determinants of dermal exposure among Nicaraguan subsistence farmers during pesticide applications with backpack sprayers. *The Annals of Occupational Hygiene* **49**, 17–24.

Buchanan, D., Pilkington, A., Sewell, C., Tannahill, S.N., Kidd, M.W., Cherrie, B. and Hurley, J.F.(2001) Estimation of cumulative exposure to organophosphate sheep dips in a study of chronic neurological heath effects among United Kingdom sheep dippers. *Occupational and Environmental Medicine* **58**, 694–701.

Buekers, J., Deutsch, F., Veldeman, N., Janssen, S. and Panis. L.I. (2014) Fine atmospheric particles from agricultural practices in Flanders: from emissions to health effects and limit values. *Outlook on Agriculture* **43**, 39–44.

Calvert, G.M., Mehler, L.N., Alsop, J., De Vries, A.L. and Besbelli, N. (2010) Surveillance of pesticide-related illness and injury in humans. In *Hayes' Handbook of Pesticide Toxicology*. Elsevier, San Diego, CA, pp. 1313–1369.

Cattani, M., Cena, K., Edwards, J. and Pisaniello, D. (2001) Potential dermal and inhalation exposure to chlorpyrifos in Australian pesticide workers. *The Annals of Occupational Hygiene* **45**, 299–308.

Chester, G. (1993) Evaluation of agricultural worker exposure to, and absorption of, pesticides. *The Annals of Occupational Hygiene* **37**, 509–523.

Chester, G. (1995) Revised guidance document for the conduct of field studies of exposure to pesticides in use. In *Methods of Pesticide Exposure Assessment*. Curry, P.B. Iyengar, S., Maloney, P.A., Maroni, M., (Eds.), pp. 179–215. NATO Challenges of Modern Society, 19. Plenum Press, New York/London.

Chester G. and Ward, R.J. (1983) An accurate method for measuring potential dermal exposure to pesticides. *Human Toxicology* **2**, 555–556.

Cinalli C., Carter, D., Clark, A. and Dixon, D. (1992) A laboratory method to determine the retention of liquids on the surface of hands. US EPA Report 747-R-92-003. Environment Protection Agency, Washington, DC.

Clay, M.M. and Clarke, S. W. (1987) Effect of nebulised aerosol size on lung deposition in patients with mild asthma. *Thorax* **42**, 190–194.

Coffman, C.W., Obendorf, S.K. and Derksen, R.C. (1999) Pesticide deposition on coveralls during vineyard applications. *Archives of Environmental Contamination and Toxicology* **37**, 273–279.

Cooper, S.E., Taylor, W.A. and Ramwell, C.T. (2006) Appraisal of sprayer washer techniques and guidance for optimal use. *Aspects of Applied Biology* **77**, 223–227.

Craig, I. and Mbevi, C. (1993) Contamination in the tropics. *Pesticides News* **19**, 3–5.

Curry, P.B., Iyengar, S., Maloney, P.A., Maroni, M. (Eds). 1995. *Methods of Pesticide Exposure Assessment*. NATO/Challenges of Modern Society, 19. Plenum Press, New York, pp. 1–224.

De Vreede, J.A.F., Brouwer, D.H., Stevenson, H. and Van Hemmen, J.J. (1998) Exposure and risk estimation for pesticides in high-volume spraying. *The Annals of Occupational Hygiene* **42**, 151–157.

Denny, R.L. (2013) Designing and implementing effective pesticide container stewardship programmes. *Outlooks on Pest Management* **24**, 244–247.

Denovan, L.A., Lu, C., Hines, C. J. and Fenske, R.A.(2000) Saliva biomonitoring of atrazine exposure among commercial herbicide applicators. *International Archives of Occupational and Environmental Health* **73**, 457–462.

Dollimore, L. and Schimpf, W. (2013) Obsolete pesticide stocks – the past 25 years, lessons learned and observations for the future. *Outlooks on Pest Management* **24**, 251–256.

Durham, W.F. and Wolfe, H.R. (1962) Measurement of the exposure of workers to pesticides. *Bulletin of the World Health Organization* **26**, 75–91.

EFSA (2014) Guidance on the assessment of exposure of operators, workers, residents and bystanders in risk assessment for plant protection products. *EFSA Journal* **12**, 55pp.

Espanhol-Soares, M., Nociti, L.A.S. and Machado-Neto, J.G. (2013) Procedures to evaluate the efficiency of protective clothing worn by operators applying pesticide. *The Annals of Occupational Hygiene* **57**, 1041–1053.

European Food Safety Authority (EFSA) (2010) Scientific opinion on preparation of a guidance document on pesticide exposure assessment for workers, operators, bystanders and residents. *EFSA Journal* **8**, 65pp.

Fenske R.A. (1990) Non-uniform dermal deposition patterns during occupational exposure to pesticides. *Archives of Environmental Contamination and Toxicology* **19**, 332–337.

Fenske, R.A. and Day, E.W. (2005) Assessment of exposure for pesticide handlers in agricultural, residential and institutional environments. In: Franklin, C.A. and Worgan, J.P. (Eds.) *Occupational and Residential Exposure Assessment for Pesticides*. Wiley, Chichester pp. 13–43.

Fernando, H.E. (1956) A new design of sprayer for reducing insecticide hazards in treating rice crop. *FAO Plant Protection Bulletin* **4**, 117–120.

Fogg, P. and Carter A.D. (1998) Biobeds: the development and evaluation of a biological system for pesticide waste and washings. *BCPC Symposium* **70**, 49–58.

Fogg, P., Boxall, A.B.A. Walker, A. and Jukes, A. (2003a) Pesticide degradation in a 'biobed' composting substrate. *Pest Management Science* **59**, 527–537.

Fogg, P., Boxall, A.B.A. Walker, A. and Jukes, A. (2003b) Degradation and leaching potential of pesticides in biobed systems. *Pest Management Science* **60**, 645–654.

Gao, B.B., Tao, C.J., Ye, J.M., Ning, J., Mei, X.D., Jiang, Z.F., Chen, S. and She, D.M. (2014) Measurement of operator exposure to chlorpyrifos. *Pest Management Science* **70**, 636–641.

Garfitt, S.J., Jones, K., Mason, H.J. and Cocker, J. (2002) Oral and dermal exposure to propetamphos: a human volunteeer study. *Toxicology Letters* **134**, 115–118.

Garthwaite, D. (2002) A survey of current farm practices in the UK. A report for the Crop Protection Association. Food and Environment Research Agency (FERA), Sands Hutton.

Gerritsen, R., Goede, H.A., Roelofs, V., Spaan, S., OOsterwijk, M.T.T., Charistou, A., Butler-Ellis, C., Glass, C.R., Machera, K., Kennedy, M., Owen, H., Scott, D. and Hart, A. (2014) The BROWSE operator exposure models for plant protection products. *Aspects of Applied Biology* **122**, 165–174.

Gilbert, A.J. and Bell, G.J. (1990) Test methods and criteria for selection of types of coveralls suitable for certain operations involving handling or applying pesticides. *Journal of Occupational Accidents* **11**, 255–268.

Glass C.R. and Mathers J.J. (2006) Novel spray test foe evaluation of materials suitable for protective coveralls for spray application. *Aspects of Applied Biology* **77** 241–246.

Glass, C.R., Cohen Gomez, E., Delgado Cobos, P., and Mathers, J.J. (1998) Modified spray test for protective clothing used in southern Europe. Proceedings of 9th International Congress Pesticide Chemistry, The Food-Environment Challenge, Westminster, London, UK, 2–7 August.

Glass, C.R., Gilbert, A.J., Mathers, J.J., Lewis, R.J., Harrington, P.M. and Perez Duran, S. (2002a) Potential for operator and environmental contamination during concentrate handling in UK agriculture. *Aspects of Applied Biology* **66**, 379–386.

Glass, C.R., Gilbert, A.J., Mathers, J.J., Martinez Vidal, J.L., Egea Gonzalez, F.J., Gonzales Pradas, E., Urena Amate, D., Fernandez Perez, M., Flores Cespedes, F., Delgado Cobos, P., Cohen Gomez, E., Moreira, J.F., Santos, J., Meuling, W., Kapetanakis, E., Goumenaki, E., Papaeliakis, M., Machera, K., Goumenou, M.P., Capri, E., Trevisan, M., Wilkins, R.M., Garratt, J.A., Tuomainen, A. and Kangas, J. (2002b) The assessment of operator, bystander and environmental exposure to pesticides. Final report EUR 20489 project contract SMT4-CT96-2048. European Commission, Brussels.

Glass, C.R., Mathers, J.J., Lewis, R.J., Harrington, P.M., Gilbert, A.J., and Smith, S. (2004) Understanding exposure to agricultural pesticide concentrates. Report to HSE unpublished.

Glass C.R., Mathers J.J., Gonzalez, F.J.E.. and Mutch, E. (2010) Evaluating the risks of pesticide exposure and suitability of mitigation measures for agricultural workers in warm climates. *Aspects of Applied Biology* **99**, 133–137.

Glass, C.R. Tsakirakis, A., Kasiotis, K., Anastasiadou, P., Charistou, A., Gerritsen-Ebben, R. and Machera, K. (2014) Pesticide dermal transfer through contact with sprayed hard surfaces to operators and agricultural workers. *Aspects of Applied Biology* **122**, 423–424.

Gustin, C.A., Moran, S.J., Fuhrman, J.D., Kurtzweil, M.L., Kronenberg, J.M., Gustafson, D.I. and Marshall, M.A. (2005) Applicator exposure to acetchlor based on biomonitoring. *Regulatory Toxicology and Pharmacology* **43**, 141–149.

Hall, R.M., Heitbrink, W.A. and Reed, L.D. (2002) Evaluation of a cab using real-time aerosol counting instrumentation. *Applied Occupational and Environmental Hygiene* **17**, 46–54.

Hamey, P.Y. (2001) The need for appropriate use information to refine pesticide user exposure assessments. *The Annals of Occupational Hygiene* **45**, S69–S79.

Harrington, P., Mathers, J., Lewis, R., Perez Duran, S. and Glass, R. (2005) Potential exposure to pesticides during amateur applications of home and garden products. In: Lichtfouse, E., Schwarzbauer, J., Robert, D. (Eds.) *Environmental Chemistry: Green Chemistry and Pollutants in Ecosystems*, vol. **26**, pp. 530–537.

Harris, D.A., Johnson, K.S. and Ogilvy, J.M.E. (1991) A system for the treatment of waste water from agrochemical production and field use. *Brighton Crop Protection Conference – Weeds*, pp. 715–722.

Honeycutt, R.C. and Day, E.W. (Eds.) (2001) *Worker Exposure to Agrochemicals – Methods for Monitoring and Assessment*. Lewis Publishers (CRC Press), Boca Raton.

HSE (2002) *Dermal Exposure Resulting from Liquid Contamination*. HSE Books, Suffolk.

Hughes, E., Zalts, A., Ojeda, J., Montserrat, J and Glass, R. (2004) Potential pesticide exposure of small-scale vegetable growers in Moreno district. *Aspects of applied Biology* **71**, 399–404.

Hughes, E.A., Zalts, A., Ojeda, J.J., Flores, A.P., Glass, R.C. and Montserrat, J.M. (2006) Analytical method for assessing potential dermal exposure to captan, using whole body dosimetry, in small vegetable production units in Argentina. *Pest Management Science* **62**, 811–818.

Johnson, G.T., Morris, S., McCluskey, J.D., Xu, P. and Harbison, R.D. (2014) Pesticide biomonitoring in Florida agricultural workers. *Occupational Diseases and Environmental Medicine* **2**, 30–38.

Jones, K.A. (2014) The recycling of empty pesticide containers: an industry example of responsible waste management. *Outlooks on Pest Management* **25**, 183–186.

Krieger, R.I. (1995) Pesticide Exposure Assessment. *Toxicological Letters* **82**, 65–72.

Krieger, R.I. and Dinoff, T.M. (2000) Malathion deposition, metabolite clearance, and cholinesterase status of date dusters and harvesters in California. *Archives of Environmental Contamination and Toxicology* **38**, 546–553.

Krieger, R.I., Bernard, C.E., Dinoff, T.M., Ross, J.H. and Williams, R.L. (2001) Biomonitoring of persons exposed to insecticides used in residences *The Annals of Occupational Hygiene* **45**, S143–S153.

Kummer, R. and van Sittert, N.J. (1986) Field studies on health effects from the application of two organophosphorous insecticide formulations by hand-held ULV to cotton. *Toxicology Letters* **33**, 7–24.

Kyriaki, R., Brennan, M., Hart, A., and Frewer, L. (2014) Pesticide risk perceptions, knowledge, and attitudes of operators, workers, and residents: a review of the literature. *Human and Ecological Risk Assessment* **20**, 1113–1138.

Landers, A. (2004) Protecting the operator – are we making an impact? *Aspects of Applied Biology* **71**, 357–364.

Lesmes-Fabian, C., García-Santos, G., Leuenberger, F. Nuyttens, D. and Binder, C.R. (2012) Dermal exposure assessment of pesticide use: the case of sprayers in potato farms in the Colombian highlands. *Science of the Total Environment* **430**, 202–208.

Llewellyn, D.M., Brazier, A., Brown, R., Cocker, J., Evans, M.L., Hampton, J., Nutley, B.P. and White, J. (1996) Occupational exposure to permethrin during its use as a public hygiene insecticide. *Annals of Occupational Hygiene* **40**, 499–509.

Lu, C., Barr, D.B., Pearson, M.A. and Waller, L.A. (2008) Dietary intake and its contribution to longitudinal organophosphorus pesticide exposure in urban/suburban children. *Environmental Health Perspectives* **116**, 537–542.

MacFarlane, E., Carey, R., Keegel, T., El-Zaemay, S. and Fritschi, L. (2013) Dermal exposure associated with occupational end use of pesticides and the role of protective measures. *Safety and Health at Work* **4**, 136–141.

Machado-Neto, J.G. (2001) Determination of safe work time and exposure control need for pesticide applicators. *Bulletin of Environmental Contamination and Toxicology* **67**, 20–26.

Machado-Neto, J.G., Matuo, T. and Matuo, Y. K. (1998) Efficiency of safety measures applied to a manual knapsack sprayer for paraquat application to maize (*Zea mays* L.). *Archives of Environmental Contamination and Toxicology* **35**, 698–701.

Machera, K. Goumenou, M., Kapetanakis, E. Kalamarakis, A. and Glass, R. (2001)Determination of potential dermal and inhalation exposure of operators, following spray applications of the fungicide penconazole in vineyards and greenhouses. *Fresenius Environmental Bulletin* **10**, 464–469.

Machera, K., Kapetanakis, E. Charistou, A., Goumenaki, E. and Glass, R. (2002) Evaluation of potential dermal exposure of pesticide spray operators in greenhouses by use of visual tracers. *Journal Environmental Science and Health* **37**, 113–121.

Margariti, M.G. and Tsatsakis, A.M. (2009) Assessment of long-term subacute exposure to dimethoate by hair analysis of dialkyl phosphates DMP and DMTP in exposed rabbits: the effects of dose, dose duration and hair colour. *Environmental Research* **109**, 821–829.

Marin Juan, A., Martinez-Vidal, J.L., Egea Gonzalez, F.J., Garrido Frenich, A. Belomonte Vega, A., Glass, C.R. and Sykes, M. (2004) Biological monitoring of greenhouse workers in Almeria. *Aspects of Applied Biology* **71**, 405–408.

Matthews, G.A (2001) Dermal exposure of hands to pesticides. In Maibach, I (Ed.) *Toxicology of Skin*. Taylor and Francis, Philadelphia, PA, pp. 179–182.

Maumbe, B.M. and Swinton, S.M. (2003) Hidden health costs of pesticide use in Zimbabwe's smallholder cotton growers. *Social Science and Medicine* **57**,1559–1571.

Mekonnen, Y. and Agonafir, T. (2002) Pesticide sprayers' knowledge, attitude and practice of pesticide use on agricultural farms of Ethiopia. *Occupational Medicine* **52**, 311–315.

Meuling, W.J.A., Franssen, A.C. Brouwer, D.H. and Van Hemmen, J.J. (1997) The influence of skin moisture on the dermal absorption of propoxur in human volunteers: consideration for biological monitoring practices. *The Science of the Total Environment* **199**, 165–172.

Moreira, J.F., Santos, J. and Glass, C.R. (1999) Personal protective equipment penetration during application of plant protection products. XIVth International Plant Protection Congress, Jerusalem, Israel, July 25–30.

Morgan M.K. and Jones, P.A. (2013) Dietary predictors of young children's exposure to current-use pesticides using urinary biomonitoring. *Food and Chemical Toxicology* **62**, 131–141.

Nelson, C. Laughlin, J. Kim, C. Rigakis, K. Mastura, R and Scholten, L. (1992) Laundering as decontamination of apparel fabrics: residues of pesticides from six chemical classes. *Archives of Environmental Contamination and Toxicology* **23**, 85–90.

Ngowi, A.V.F., Maeda, D.N. and Partanen, T.J. (2001a) Assessment of the ability of health care providers to treat and prevent adverse health effects of pesticides in agricultural areas of Tanzania. *International Journal of Occupational Medicine and Environmental Health* **14**, 349–356.

Ngowi A.V.F., Maeda, D.N., Partanen, T.J., Sanga, M.P. and Mbise, G. (2001b) Acute health effects of organophosphorus pesticides in Tanzanian small-scale coffee growers. *Journal of Exposure Analysis and Environmental Epidemiology* **11**, 335–339.

Nigg, H.N., Stamper, J.H. and Mallory, L.L. (1993) Quantification of human exposure to ethion using saliva. *Chemosphere* **26**, 897–906.

Nutley, B.P., Berry, H.F., Roff, M., Brown, R.H., Niven, K.J.M. and Robertson, A. (1995) The assessment of operator risk from sheep dipping operations using organophosphate based dips. In: Best, G. and Ruthven, D. (Eds.) *Pesticides: Developments, Impacts and Controls*. Royal Society of Chemistry, London, pp. 43–53.

Nuyttens, D., Windey, S., Braekman, P., de Moor, A. and Sonck, B. (2004) Comparison of operator exposure for five different greenhouse spraying applications. *Aspects of Applied Biology* **71**, 349–356.

Nuyttens, D., Braekman, P., Windey, S. and Sonck, B. (2009) Potential dermal pesticide exposure affected by greenhouse spray application technique. *Pest Management Science* **65**, 781–790.

Ohayo-Mitoko, G.J.A., Kromhout, H. Karumba, P.N. and Boleij, J.S.M. (1999) Identification of determinants of pesticide exposure among Kenyan agricultural workers using empirical modelling. *Annals Occupational Hygiene* **43**, 519–525.

Rajan-Sithamparanadarajah, R., Roff, M., Delgado, P., Eriksson, K., Fransman, W., Gijsbers, J.H.J., et al. (2004). Patterns of dermal exposure to hazardous substances in European Union workplaces. *The Annals of Occupational Hygiene* **48**, 285–297.

Ramos, L.M., Querejeta, G.A., Flores, A. P., Hughes, E.A., Zalts, A. and Montserrat, J.M. (2010) Potential Dermal Exposure in greenhouses for manual sprayers: Analysis of the mix/load, application and re-entry stages. *Science of The Total Environment* **408**, 4062–4068.

Ramwell, C.T., Johnson, P.D., Boxall, A.B.A. and Rimmer, D.A. (2004) Pesticide residues on the external surfaces of field-crop sprayers: Environmental impact. *Pest Management Science* **60**, 795–802.

Ramwell, C.T., Cooper, S.E. and Taylor W.A. (2006) The environmental impact of cleaning the external surfaces of sprayers. *Aspects of Applied Biology* **77**, 229–233.

Rando, J.C.M. (2013) The campo limpo system reverse logistics for empty containers of crop protection products. *Outlooks on Pest Management* **24**, 273–275.

Renwick, A.G. (1991) Safety factors and establishment of acceptable daily intake. *Food additives and Contaminants* **8**, 135–150.

Roff, M.W. (1994) A novel lighting system for the measurement of dermal exposure using a fluorescent dye and an image processor. *The Annals of Occupational Hygiene* **38**, 903–919.

Ross, J.H., Driver, J.H., Cochran, R.C., Thongsinthusak, T. and Krieger, R.I. (2001) Could pesticide toxicology studies be more relevant to occupational risk assessment? *The Annals of Occupational Hygiene* **1001**, S5–S17.

Ross, J., Driver, J., Lunchick, C., O'Mahony, C. (2015) Models for estimating human exposure to pesticides. *Outlooks on Pest Management* **26**, 33–37.

Shaw, A., Lin, Y.J. and Pfell, E. (1996) Effect of abrasion on protective properties of polyester and cotton/polyester blend fabrics. *Bulletin of Environmental Contamination and Toxicology* **56**, 935–941.

Shaw, A., Lin, Y.J. and Pfell, E. (1997) Qualitative and quantitative analysis of diazinon in fabric exposed to various simulated sunlight and humidity conditions. *Bulletin of Environmental Contamination and Toxicology* **59**, 389–395.

Shaw, A., Nomula, R. and Patel, B. (2000) Protective clothing and application controls for pesticide application in India: a field study. In Nelson, C.N. and Henry, N.W. (Eds.). *Performance of Protective Clothing: Issues and Priorities for the 21st Century*. Vol. **7**. ASTM STP 1386. ASTM, West Conshohocken, PA.

Stephens, T., Woods, S., Cooper, S.E. and Goddard, R.S. (2006) Considering the use of air induction nozzles to reduce operator contamination when using knapsack sprayers. *Aspects of Applied Biology* **77**, 235–240.

Sutherland, J.A., King, W.J., Dobson, H.M. Ingram, W.R., Attique, M.R. and Sanjrani, W. (1990) Effect of application volume and method on spray operator contamination by insecticide during cotton spraying. *Crop Protection* **9**, 343–350.

Thornhill, E.W., Matthews, G.A. and Clayton, J.S. (1996) Potential operator exposure to insecticides: a comparison between knapsack and CDA spinning disc sprayers. *Proceedings of the Brighton Conference*, Vol. **3**, BCPC, Farnham, UK, pp. 1175–1180.

Tsakirakis, A.N., Kasiotis, K.M., Charistou, A.M., Arapaki, N., Tsatsakis, A., Tsakalof, A. and Machera, K. (2014) Dermal & inhalation exposure of operators during fungicide application in vineyards. Evaluation of coverall performance. *Science of the Total Environment* 470–471, 282–289.

Tsatsakis, A.M., Barbounisa, M.G., Kavalakisa, M., Kokkinakisa, M., Terzia, I. and Tzatzarakisa, M.N. (2010) Determination of dialkyl phosphates in human hair for the biomonitoring of exposure to organophosphate pesticides. *Journal of Chromatography B* **878**, 1246–1252.

Tsatsakis, A.M., Christakis-Hampsas, M. and Liesivuori, J. (2012) The increasing significance of biomonitoring for pesticides and organic pollutants. *Toxicology Letters* **210**, 107–109.

Tunstall, J.P. and Matthews, G.A. (1965) Contamination hazards in using knapsack sprayers. *Empire Cotton Growing Review* **42**, 193–196.

Van Drooge, H.L., Groeneveld, C.N. and Schipper, H.J. (2001) Data on application frequency of pesticide for risk assessment purposes. *The Annals of Occupational Hygiene* **45**, S95–S101.

Van Hemmen, J.J. (2001) EUROPOEM, a predictive occupational exposure database for registration purposes of pesticides. *Applied Occupational and Environmental Hygiene* **16**, 246–250.

Van Hemmen, J.J. and Brouwer, D.H. (1997) Exposure assessment for pesticides: operators and harvesters risk evaluation and risk management. *Med. Fac. Landbouww. University of Gent* **62**(2), 113–130.

Van Hemmen, J.J. and Van der Jaagt, K.E. (2005) Generic Operator Exposure Databases. In Franklin, C.A. and Worgan, J.P. (Eds.) *Occupational and Residential Exposure Assessment for Pesticides*. Wiley, Chichester.

Wang, J., Timchalk, C. and Lin, Y. (2008) Carbon nanotube-based electrochemical sensor for assay of salivary cholinesterase enzyme activity: an exposure biomarker of organophosphate pesticides and nerve agents. *Environmental Science and Technology* **42**, 2688–2693.

Wester, R.C., Sedik, L., Melenderes, J., Logan, F. Maibach, H.I and Russel, I. (1993) Percutaneous absorption of diazinon in humans. *Food Chemistry and Toxicology* **31**, 569–572.

Wicke, H., Backer, G. and Friéleben R. (1999) Comparison of spray operator exposure during orchard spraying with hand-held equipment fitted with standard and air injector nozzles. *Crop Protection* **18**, 509–516.

Williams, R.L., Aston, L.S., Krieger, R.I. (2004). Perspiration increased human pesticide absorption following surface contact during an indoor scripted activity program. *Journal of Exposure Analysis and Environmental Epidemiology* **14**, 129–136.

Wolf, T.M., Gallender, K.S., Downer, R.A., Hall, F.R., Fraley, F.W. and Pompeo, M.P. (1999) Contribution of aerosols generated during mixing and loading of pesticides to operator inhalation exposure. *Toxicology Letters* **105**, 31–38.

Wolfe, H.R. (1976) Field exposure to airborne particles. In Lee, R.E. (Ed.) *Air Pollution from Pesticides*. CRC Press, Cleveland, pp. 137–161.

Yuantari, M.G.C., Van Gestel, C.A.M., Van Straalen, N.M., Widianarko, B., Sunoko, H.R. and Shobib, M.N. (2015) Knowledge, attitude, and practice of Indonesian farmers regarding the use of personal protective equipment against pesticide exposure. *Environmental Monitoring and Assessment* **187**, 142.

5 Spray drift, bystander, resident and worker exposure

Following the ban on organochlorine insecticides due to their persistence in the environment, there has continued to be considerable criticism about the use of pesticides in general adversely affecting our environment. As noted earlier this has led to many changes in the regulation of pesticides in Europe. The concern has focussed on the movement of pesticides out of the agricultural fields treated to protect crops by 'spray drift' and contamination of water. In the UK, following concerns about alleged health issues, the government asked the Royal Commission on Environmental Pollution to report on *Crop Spraying and the Health of Residents and Bystanders* (RCEP, 2005). Following this report and a Judicial Review (Warren, 2009), a research project to develop a 'Bystander and resident exposure assessment model' (BREAM) was set up (Kennedy et al., 2012).

Later, the European Union set up a project to review, improve and extend the models currently used in the risk assessment of plant protection products (PPPs) to evaluate bystander, residents, operators and workers exposure to pesticides (BROWSE). The aim was to use improved exposure models as part of the implementation of Regulation 1107/2009 on authorisation of PPPs, which replaced Directive 91/414/EC. In addition to assessing downwind drift, further consideration was given to touching surfaces on which spray was deposited. As part of the BROWSE project, and implementation of the sustainable use of pesticides (SUD), a consultation process was implemented, which involved relevant stakeholders to answer an electronic questionnaire to identify their opinions regarding the factors influencing pesticide exposure which needed to be prioritised. It also identified the most suitable formats to develop training and awareness of target groups. Sacchettini et al. (2012) considered that the data collected were in line with a 'doctrine of engagement' (Walls et al., 2010) that recognised the benefits of the involvement of stakeholders in the policy

Pesticides: Health, Safety and the Environment, Second Edition. G.A. Matthews.

process in terms of political efficacy, gaining of trust in policy makers. Similarly Remoundou et al. (2014) review risk perceptions, knowledge levels and attitudes of operators, workers, and residents in relation to non-dietary exposure to agricultural pesticides to assist policy-makers, and risk communicators in the development of targeted training and awareness-raising material.

This chapter considers the movement of pesticides from treated crops and the extent to which bystanders and people living near agricultural areas are exposed to spray drift. The exposure of those working in crops that have been sprayed, namely farm workers, who come into contact with pesticide deposits is also considered.

The downwind movement of spray droplets by the wind at the time of a spray application to areas, which are outside the treated fields, can cause unacceptable effects depending on the type of pesticide. Insecticides may adversely affect bees and other beneficial or non-target insects, while herbicides can affect vegetation, which can in turn cause a reduction in non-target species by the effects on habitats and food sources. The general public have also felt that airborne spray is the cause of many illnesses. In the UK, any specific complaint made by a member of the public, related to agricultural use of pesticides, excluding sheep treatments covered by the Medicines Act 1968, is examined carefully by the Pesticide Incident Appraisal Panel (PIAP), which is administered by the HSE. There were 204 complaints, 62 of which alleged ill health, while 142 were related to environmental concerns in 2003–2004 (Anon, 2004). Since then the number of reports each year has decreased. Complaints related to boom sprayers were due to their use in windy conditions. Only in a small proportion of the incidents reported is a definite occurrence of spray drift confirmed as a possible cause of the complaint. In many cases complaints have been due to detection of the odour either of the active ingredient of the pesticide but also in some cases due to the solvent or carrier used in the formulation, and likely to be vapour drift rather than movement of spray droplets. This is due to the sensitivity of the human nose to certain odours at extremely low concentrations.

The system of reporting pesticide incidents to HSE was criticised as it dealt almost entirely with acute symptoms and very few cases of chronic illness were reported. Although the Royal Commission on the Environment recommended PIAP should be part of the Health Protection Agency under the Department of Health rather than the Health and Safety Executive (HSE), it has continued under the HSE and has issued guidelines to raise awareness of PIAP and ensure the HSE complaints procedures are followed. (see http://www.hse.gov.uk/foi/internalops/sims/ag_food/010805.htm). Trials to assess the feasibility of post-event investigation of incidents of spray drift reported by members of the public were reported by Rimmer et al. (2006).

Downwind drift of vapour may also occur after a spray application from foliar deposits due to the volatility of the pesticide, if temperatures are sufficiently high. Widespread serious phytotoxic effects on some broad–leaved vegetable crops occurred in the 1970s in England, when there was exceptionally warm weather after a volatile herbicide had been applied on cereals (Elliot and Wilson, 1983). This led to improvements in the formulation of the herbicide to reduce volatility. In Canada, drift potential of the butyl ester was 8–10 times greater than that from the dimethylamine formulation of 2,4-D in trials where 25–30% of the butyl ester formulation was collected as vapour drift in the 1/2h after spraying (Grover et al., 1972). Vapour drift can occur more than 12h after application, especially if temperatures are high. Although the manufacturer has developed a new formulation of 2,4-D for use with crops genetically engineered to be 2,4-D tolerant, concern about its use have been raised. This is partly due to confusion with the infamous 'Agent Orange', which also contained 2,4,5-T contaminated with dioxins and was highly toxic.

What is drift?

Himel (1974) defined spray drift into two categories, namely exo- and endo-drift (Fig. 5.1).

Exo-drift

This is the movement of spray droplets beyond the edge of a treated field. Many studies have measured deposition of spray droplets on the ground or other horizontal surfaces over a relatively short distance downwind of a treated field. However more attention has now been given to the much

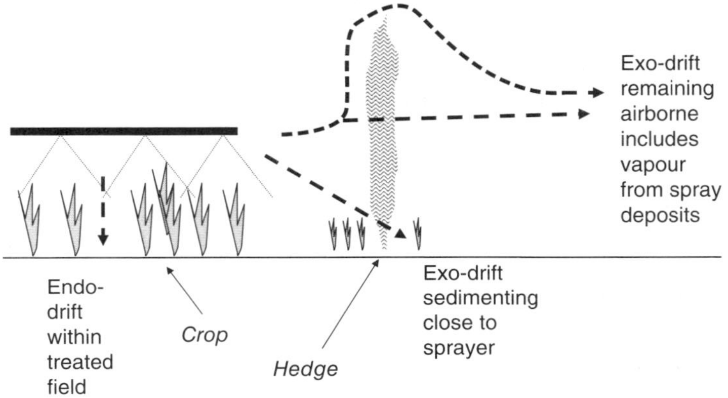

Fig. 5.1 Exo- and endo-drift.

smaller volume of an agricultural spray made up of the smallest droplets which will tend to remain airborne. Depending on the movement of air, these droplets can travel over very long distances, especially if the meteorological conditions including convective pockets of air favour their movement upwards, before they ultimately sediment on vegetation or on the ground. According to Jensen and Olesen (2014) airborne drift could be substantial, based on an examination of the difference between the amount calculated from sampling spray deposited on the crop and ground using tunnel sprayers in vineyards and a study of the application at different times of day in apples. Their analysis indicated that large fractions of the spray were not detected using traditional orchard sprayers in vineyards and orchards.

Endo-drift

This is the distribution of a pesticide within a field, but not on the intended target. Thus some of a foliar spray may fail to be deposited on leaves and sediment on the ground. Rain soon after a spray application may also wash deposits off foliage on to the ground. Subsequent movement of this pesticide deposit by leaching through the soil profile can lead to contamination of ground water, if the pesticide molecule is sufficiently mobile in soil and not adsorbed on soil particles. This endo-drift could also become exo-drift, if heavy rain or irrigation washed the soil surface deposit into the nearest ditch or waterway. Although not spray drift in the strict sense of downwind movement of droplets, another point source of pesticide contamination of surface water or drains is the water used to wash a sprayer or if there is any spillage from a pesticide container.

Peak levels of pesticide in river water samples, caused by surface runoff following storms involving 6.8–18.4 mm rain/day, were from 2 to 41 times higher than the levels recorded due to spray drift (0.04–0.07 μg/l), even when the river discharge rate was much greater 7–22.4 m^3/s instead of 0.28 m^3/s (Schulz, 2001). In Germany 24 g pesticides were found in each farmyard runoff during the application period, presumably caused by cleaning the spraying equipment (Neumann et al., 2002). This has led to the need for an additional water tank on the sprayer so at least the inside of the tank is washed in the field and the washings used to spray the last part of a field.

How is drift measured?

An international standard for measurement of drift (ISO 22866) is designed to cover both short-range down wind sedimentation of spray and longer range airborne drift of the smaller spray droplets.

Droplets that sediment

Most attention has been given to the spray that is collected at ground level (Fig. 5.2). Trials have been concerned with the distribution of spray obtained with different types of equipment, but for registration requirements, the sedimentation data has been needed in risk assessments of drift on to surface water and has led to the adoption of buffer zones. The trajectory of droplets larger than 200 μm from a tractor boom is influenced primarily by gravity, so fall out on a horizontal surface occurs mostly within about 5 m from the end of the boom, hence this basic distance is used for buffer zones to protect water-courses.

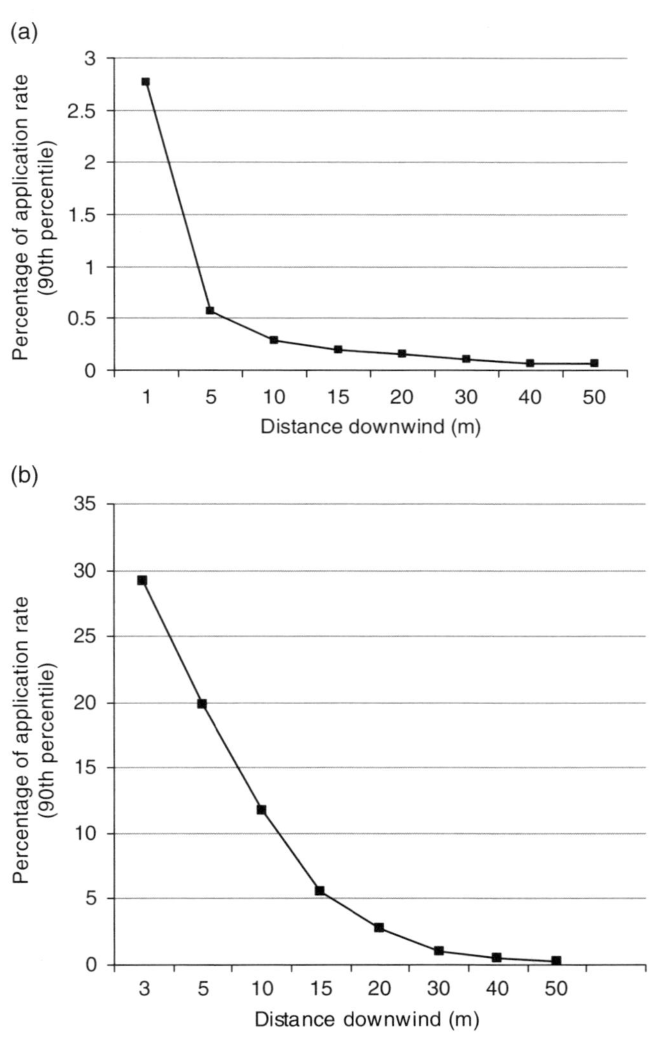

Fig. 5.2 Example of drift measurements from downwind of a spray boom (a) drift from field crop and (b) drift from orchard early season.

Different sampling methods have been used, but most have relied on sampling spray deposited on a horizontal surface. Measurements have been made when applying pesticides that are easily measured, but most assessments are made with a tracer dye quantified by spectro-fluorometric analysis. The dye used should be non-toxic and relatively inexpensive. The food colourant tartrazine is one dye that has been used. A dye also must be photo-stable as the samples cannot all be collected simultaneously. Gil and Sinfort (2005) considered that brilliant sulphoflavine was the best fluorescent dye due to its low degradation in sunlight. Some drift studies have used EDTA chelates of metals as spray tracers, as this allows spraying with different equipment over the same target area to be done and then separate the deposits of each tracer on the same samples (Cross et al., 2001; Foque et al., 2014). Neutron activation analysis, using dysposium chloride added to the spray, has also been used to detect drift (Dobson et al., 1983). A combination of a fluorescent dye and pesticide is sometimes used to allow a large number of samples to be examined qualitatively and restricting chemical analyses to sub samples that relate to different levels of spray coverage (Matthews and Johnstone, 1968).

Chromatography paper is often used to collect spray, but glass surfaces have the advantage that the deposit of a pesticide or tracer can be easily washed off and measured quantitatively. Petri dishes have been used but these suffer from a raised edge that can affect the air-flow and thus movement of smaller droplets close to the dish. Flat glass plates avoid this problem, but these must be kept exactly horizontal during sampling, as deposition can be influenced even if the plates are placed at a slight angle. In studying the distribution of spray across the swath of an aircraft, Johnstone and Matthews (1965) used table tennis balls mounted on pins to sample the spray containing a red dye tracer to avoid the difficulty of having plates set correctly in a grass field.

The amount of spray deposited by sedimentation will decline as distances downwind increase so the sample size needs to be sufficiently large to collect a measurable amount of the tracer. An individual sample of 100 cm^2 is usually considered necessary. The ISO standard for drift measurement also requires a minimum total area of 1000 cm^2 to be sampled. Apart from having samples at different distances down wind, ideally the sampling line is replicated to allow for variations in wind speed and direction while taking the sample. Often costs prevent adequate number of samples being taken. Visual assessments of spray drift, at least over short distances can be done by setting out small strips of water-sensitive paper that changes colour where water droplets are deposited on it.

The cost of doing field assessments of spray drift and variability due to changes in wind velocity and direction have led to efforts to assess the potential spray drift from different nozzles in a wind tunnel (Miller et al., 1993; Nuyttens et al., 2009; Douzals and Heidary, 2014) (Fig. 5.3).

Fig. 5.3 Measuring potential drift under wind tunnel conditions (Silsoe Research Institute).

Airborne droplets

The droplets that remain airborne are generally smaller than 100 μm and unless the liquid is involatile, the droplets will become smaller in flight. Efficient sampling of these small droplets is more difficult as they can bypass a solid object in their path, when carried in the airflow. Passive samplers, such as 2 mm diameter polythene line suspended have acceptable collection efficiency, dye can be recovered and quantified at reasonable cost (Gilbert and Bell, 1988; Miller, 1993). Mathers et al. (2006) considered that an array of lines to measure the vertical drift profile may provide the best artificial target to estimate bystander exposure during field trials (Fig. 5.4).

Such lines are less expensive to use than samplers that draw air through a filter at a calibrated rate. However, washing the filter allows the amount of tracer to be quantified in relation to the volume of air sampled and the sampling time. Small units can be attached to a person to sample air near the face (breathing zone) to assess the risk of inhalation of a pesticide. Cascade impactors are a specialised high-volume air sampler that separates

Fig. 5.4 Sampling airborne drift of spray droplets (Silsoe Research Institute).

samples in relation to the size of droplets. Pumps to draw air through the sampler require a source of power, so field samples are often taken with rotary samplers, such as rotorods, which can be battery operated.

Lidar measurements have been made to assess long-range drift to validate models such as the USDA Forest Service Cramer-Barry-Grim (FSCBG). Stoughton et al. (1997) reported that aerial sprays over a forest canopy drifted more than 2000 m from the spray line, thus further than predicted during near-neutral conditions (Miller and Stoughton, 2000).

Briand et al. (2002) compared different samplers, in conjunction with gas chromatography-mass spectrometry, to evaluate spray drift from an orchard. Compared to earlier studies, the relatively high concentrations that were detected in the gas phase, indicated that evaporation in the high temperatures from small droplets allowed them to drift. Thus temperature and relative humidity as well as the physical properties of the pesticide will influence the vapour and particle distribution, making it important to differentiate between drift and post application transfer from deposits.

The time-weighted air concentration of pesticides downwind of a barley field due to spray drift in Germany, was highest in the first 2 h after application, then decreased. Over 21 h, up to 0.58 μg/m^3 was detected at 10 m downwind. Volatile insecticides were detected at up to 200 m in the first 2 h, but below the limit of quantification (Siebers et al., 2003).

Few studies have sampled air in the UK for pesticides; but where these have been done (Turnbull, 1995), the highest quantities observed over 24 or 48 h were in samples taken near to field applications. One sample was just over 2000 pg/m^3, which is 42,000 times less than the air concentration measured at the bystander position 8 m from the boom. Mean values of pesticides in air were generally less than 400 pg/m^3. Similar air quality studies elsewhere have generally detected the most volatile of pesticides, such as those, like methyl bromide, used to fumigate soil (Lee et al., 2002). The amount of a pesticide in the air declines due to dilution with other air, removal by rain and degradation of the pesticide. Restricting use of organochlorine insecticides has led to a 10-fold decrease in gamma HCH (lindane) deposits in the atmosphere in France (Teil et al., 2004). In contrast, air samples in Birmingham showed no decline in gamma HCH, but DDT concentrations were much lower in 1999/2000 compared to 1997/1998.(Harrad and Mao, 2004).

Rainfall will wash airborne spray droplets from the atmosphere. In a study in Iowa, concentrations of herbicides detected in rain varied from non-detectable amounts to as much as 154 µg/l, the latter occurring when atrazine had been applied on an extremely humid day immediately followed by <10 mm rain. This only occurred locally with other collectors detecting 1.7 µg/l. Total annual samples in rain represented less than 0.1% of the amount applied (Hatfield et al., 1996). Similar data from Germany confirms that atrazine was detected in rain (Epple et al., 2002). Other data of amounts of pesticide in rainwater, which are generally below µg/l are given by Unsworth et al. (1999) and Van Dijk and Guicherit (1999), who reported on long-range transport of pesticides in the atmosphere to IUPAC Commission on Agrochemicals and the Environment. A review was also published by the EU Sanco (2008).

Bystander exposure

A bystander is a person who is located within or adjacent to an area where pesticides are being applied or has just been applied, but whose presence is quite incidental and unrelated to the application of the pesticide. There is a risk that some of the spray drifting downwind could be deposited on their body, especially if no action is taken to avoid exposure (Fig. 5.5). A bystander, who happens to be walking near a field while it is being sprayed, would not be wearing any personal protective clothing (PPE). Workers, who enter a field, for example for harvesting a crop are considered separately below. However in addition to people walking near fields being sprayed, concern has been expressed about spray drifting across field boundaries into adjacent gardens and houses, occupied by 'residential neighbours'.

When the Advisory Committee on Pesticides (ACP) considered exposure of residents to sprays during the cropping season from nearby fields, it concluded that the approach, then used, was protective of long-term bystander exposure. However the Royal Commission on Environmental Pollution considered that bystander exposure methodology was conservative and protective in toxicological terms, but it was limited by using the mean of measurements taken on a person standing at one distance downwind of a sprayer. As people vary in their sensitivity to chemicals (Fig. 5.6), they suggested that risk management

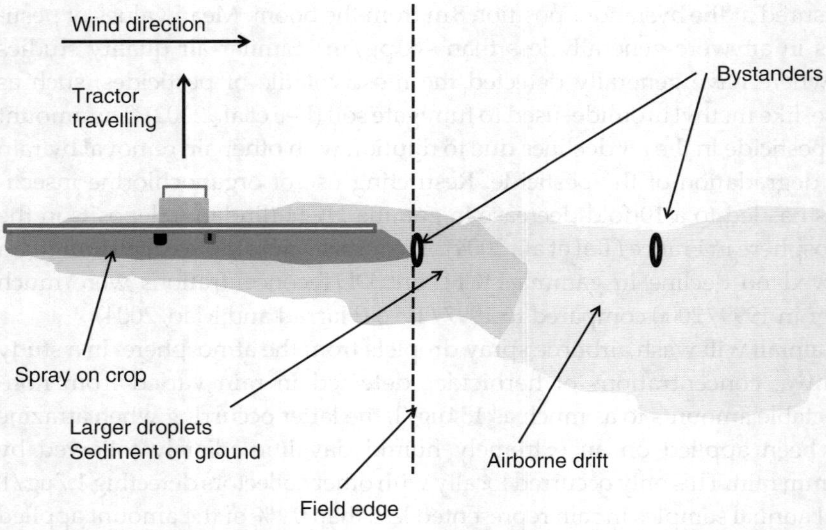

Fig. 5.5 Differences in sensitivity of individuals in relation to distribution of bystander exposure. Adapted from a diagram by Professor Brian Hoskins, Royal Commission on Environmental Pollution).

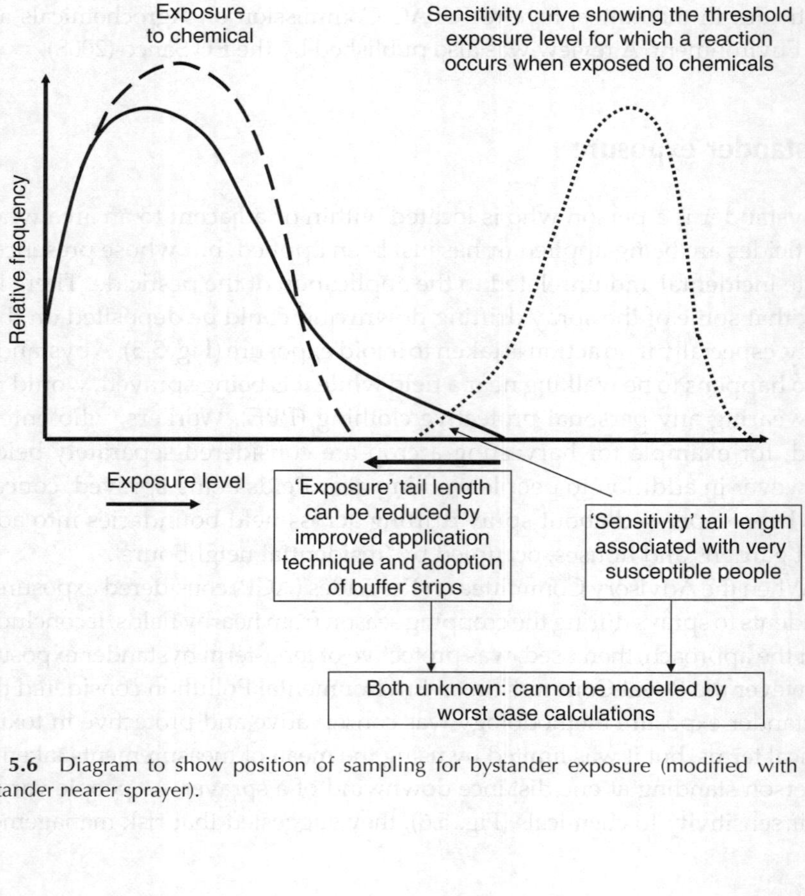

Fig. 5.6 Diagram to show position of sampling for bystander exposure (modified with one bystander nearer sprayer).

urgently needed probabilistic models to take account of worst-case situations. They also felt that people should be able to know what chemicals have been sprayed, as such information is important when examining any person who may be ill as a result of exposure to the pesticide.

In the UK, exposure of bystanders had been based on data from Gilbert and Bell (1988), who used tracer studies to measure the amount of airborne tracer that reached the breathing zone and the spray deposited on the clothing and uncovered skin of a person standing at 8 m downwind from the edge of an area treated with a tractor sprayer with a horizontal boom.

In the regulatory risk assessment of exposure to bystanders, it is now considered that the distance of a sprayer from the field boundary and the position of the bystander or resident should be 2 m distance rather than the 8 m previously used (Figs. 5.7 and 5.8). The distance of 8 m had been chosen as any person closer would be more likely to have some involvement in the pesticide application, and therefore be wearing at least overalls. Recent studies (Butler-Ellis et al., 2010a) had shown that exposures of bystanders can be significantly higher than those measured by Lloyd and Bell (1983).

Fig. 5.7 Using a colour dye to measure exposure to bystanders at two distances downwind of the end of a spray boom (Central Science Laboratory).

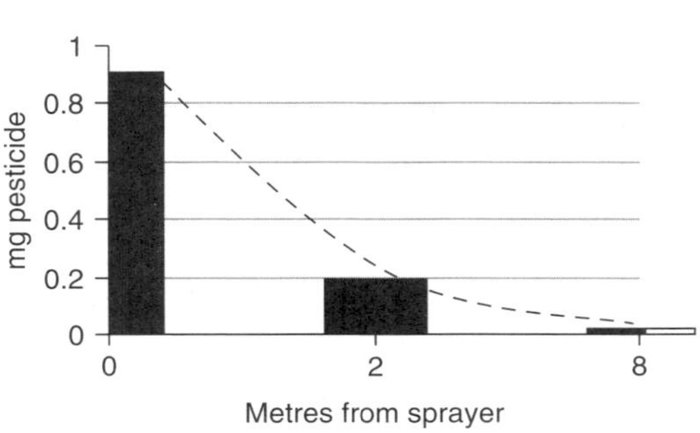

Fig. 5.8 Decrease in exposure at different distances downwind of the boom.

Table 5.1 Volume (ml per pass) of spray deposit on bystanders with two nozzles measured at two distances from the sprayed area, and mean wind speed measured at 2.0 m above the ground (Adapted from Butler-Ellis et al., 2010a)

Hypro F110 03 at 300 kPa					Billericay Bubblejet 025 at 300 kPa				
Wind speed	Child (Mannequin)		Adult (volunteer)		Wind speed	Child (Mannequin)		Adult (volunteer)	
m/s	2 m	5 m	2 m	5 m	m/s	2 m	5 m	2 m	5 m
4.95	0.25	0.14	0.41	0.24	8.18	0.29	0.18	0.41	0.38
5.86	0.26	0.11	0.58	0.29	7.14	0.19	0.13	0.41	0.27
4.16	0.17	0.06	0.25	0.12	6.04	0.24	0.09	0.56	0.19
3.55	0.25	0.10	0.46	0.20	5.43	0.24	0.05	0.39	0.15

It is also assumed that a bystander may spend 30 min on a footpath across a treated crop to reflect a worse-case scenario, although the EFSA default value is 15 min.

In carrying out these trials, differences in wind speed were taken into account in estimating potential exposure of the bystander under a range of weather conditions. Much depends on the choice of nozzle, operating pressure and the height of the boom as well the filtering effect of the crop.

An example of data showing exposure in ml/pass for a child and adult at 2 and 5 m downwind from the sprayed area shows that less spray is deposited at 5 m and lower wind speeds (Table 5.1). Despite the higher wind speeds during tests, bystanders were less exposed when the air induction nozzle (Bubblejet) was tested – compare 5.86 with 6.04 m/s wind speeds, a difference that would increase further from the sprayer. This would be expected as the air induction nozzle produces a lower proportion of droplets smaller than 100 μm, which can remain airborne (Piggott and Matthews, 1999).

On average, a bystander was exposed to 0.32 ml of spray for a single pass of the sprayer. A threefold increase in the total deposition on a bystander, who stayed in the same position, was estimated to account for any spray that may reach a bystander as the sprayer moves further away upwind. However, more of the spray would be deposited on the crop, especially as the smallest airborne droplets tend to be carried around a solid body. These trials did not measure inhalable droplets as previous work has shown that the measurements were below the limit of quantification.

Previous work had indicated that a mean potential inhalation exposure of 0.02 ml/m^3 translates into an inhalation exposure of 0.006 ml spray assuming an exposure period of 5 min, a respiratory rate of 3.6 m^3/h and all the pesticide was absorbed. Inevitably under field conditions with variations in wind speed and direction, there will be some variability in measurements. The adoption of a mean value has been criticised with the suggestion that more account should be taken of the probability of more exposure under some conditions.

Butler-Ellis and Miller (2010) updated an earlier computer model of spray drift (Miller and Hadfield, 1989) to predict non-target exposures to pesticides, specifically to improve estimates of bystander dermal and inhaled spray drift exposure. The model takes into account the effects of multiple fan nozzles (FF/110/1.2/3.0) on a boom and a faster forward speed to simulate application conditions appropriate to current field practice, particularly in the UK. Another model has examined the dispersion of volatilised pesticides in air after application for the assessment of resident and bystander exposure (Butler Ellis et al., 2010b). This model separates the emission of pesticide vapour from its downwind dispersion, allowing the latter to be tailored to UK conditions by using representative meteorological data. However, they pointed out that further studies were needed to develop an appropriate model of emissions that can take into account environmental conditions and physico-chemical properties of the formulation.

Bystander exposure to sprays in orchards was first evaluated by Lloyd et al. (1987), who showed that a person was exposed to 3.7 ml of spray at 8 m downwind with 0.002 ml/person in the breathing zone. The increased exposure was related to the spray being directed up into trees, so more droplets are airborne in contrast to the arable situation with downwardly directed sprays. Although new methods of spraying trees have been introduced, especially where tree canopies are not so large, changes in spraying techniques with air-assisted sprayers have not changed so much. Van de Zande et al. (2014) compared the spray drift from a cross-flow fan sprayer and an axial fan at different stages (dormant, intermediate and full leaf) of growth in an orchard. In examining the downwind movement of spray from orchards, comparable results from using scourer balls and horizontal polythene lines were obtained (Butler-Ellis et al., 2014), but there was considerable variability in the downwind movement of spray. Bystanders were more exposed when the wind direction was along tree rows, but less when air induction nozzles were used. Drift was also reduced by using multiple-row sprayers in which the nozzles and air flow were position on vertical units between tree rows (Wenneker et al., 2014).

For the bystander, larger droplets still airborne over a short distance downwind from the sprayer may impact on parts of the body, but the smallest will tend to be carried in the air-flow around the body. Droplets in the size range that can be inhaled (<10 μm) are those most likely to be carried by wind away from a bystander. Clearly, the coarser the spray, there is less exposure to the bystander, which is confirmed when air induction nozzles, which produce only a small volume of spray in droplets below 100 μm diameter (typically <5%), were used.

Generally in open fields, inhalation exposure is negligible due to the extremely small volume of spray in droplets small enough to be inhaled into the lungs. Nevertheless there is concern about downwind spray drift, which may be deposited on bystanders, as this is the cause of some reports to the Health and Safety Executive. However, tests have shown that generally the exposure of unprotected bystanders is only a fraction compared with the spray operator (Gilbert and Bell, 1988). In practice people entering fields after a pesticide treatment may be aware of an odour, which may be due to the pesticide or a volatile chemical used in the formulation. Exposure to the deposits, once they have dried will be depend on how dislodgeable the spray deposit is when they are in contact with it. There is also a risk of vapour from spray deposits, which is likely to be highest immediately after an application.

In assessing registration of pesticide products, the EFSA guidelines have suggested dermal and inhalation exposures in terms of the diluted spray (Table 5.2).

In a study in Washington, USA, high-volume air samplers were used to detect the organophosphate insecticide methamidophos applied by aircraft to five fields of potato crops located around a residential community. Data from samples were taken 12 h before, during and 24 h after the application were compared with a predictive model, which tended to under-estimate immediately after the spray but over-estimate the emission the next day (Ramaprasad et al., 2004). In modelling the drift from these aerial sprays, Tsai et al. (2005) used the US Environmental Protection Agency's Fugitive Dust Model (FDM) for its flexibility in terms of both inputs and outputs. Nevertheless, the data demonstrated the potentially high risk of inhalation exposure after the spraying was completed, when a volatile pesticide has been aerially applied. The importance of pesticide volatility was stressed by Scheyer et al. (2005) who showed that in an urban environment insecticides such as gamma HCH can be detected at between 0.01 and 1 ng/m^3 following soil–air transfer and that endosulfan was also present in 2003 in the summer months in eastern France.

In the USA, the aim of the Federal Clean Air Standards for ambient air quality is to keep volatile organic content (VOC) of pesticide products capped at 20%. In California, out of 787 pesticide products that were re-evaluated, only 372 were still registered in 2012. The problem is mainly associated with fumigants applied to soil or in storage buildings. Within Europe, apart from concern about the toxicity of the herbicide trifluralin to fish, it did

Table 5.2 EFSA guideline on dermal and inhalation exposure of bystanders

Method of application	Dermal exposure: Adults	Dermal exposure: Children	Inhalation exposure: Adults	Inhalation exposure: Children
Arable/ground boom applications	5.33 ml spray dilution/person	5.33 ml spray dilution/person	0.022 ml spray dilution/person	0.022 ml spray dilution/person
Orchard/broadcast air-assisted applications	285 ml spray dilution/person	285 ml spray dilution/person	0.00435 ml spray dilution/person	0.00435 ml spray dilution/person

not meet the requirements for inclusion in Annex I due to its high volatility and transport through air, despite a rapid photochemical degradation.

The risk assessment in the UK assumes bystanders are exposed at the same daily level for three months, which is considered far longer than those living next door to a treated field would actually experience. Furthermore, the highly toxic pesticides such as methamidophos are not approved in the UK and with changes in formulation to reduce volatility and improvements in application technology to minimise spray drift the potential for exposure to bystanders in the vicinity of treated fields has been reduced.

Residential exposure

There has been considerable concern raised by people living in houses close to farmland, who have alleged ill health due to pesticides used on neighbouring fields, drifting on to their property (Fig. 5.9). However, in addition to any spray that may drift into these residential properties, pesticides are now also used in homes and gardens, both in rural and urban areas. The extent to which homes are treated will vary significantly, but in warmer climates and where buildings have significant timber construction (structural pest control), insecticides are applied for vectors of disease, particularly mosquitoes and flies, household pests, such as cockroaches, termites, ants and other pests. Garden use is often mostly herbicides used on lawns, but fungicides and insecticides are also used to protect flowers and some vegetables. It has been estimated that 90% of all US households use pesticides (Driver and Whitmyre, 1996). In the UK only a limited number of pesticides are registered for non-professional use and some of these are now marketed primarily in ready-to-use formulations packed in trigger-operated hand sprayers or pressure packs. As in many countries these products are readily available in garden centres and supermarkets.

In the UK, there is little information about how household pesticides are used. In one study involving a sample of 147 parents, despite safety being an important issue, a third said they would not follow the label instructions exactly and almost 50% considered the labels were inadequate and difficult to understand (Grey et al., 2004), so very few were taking notice of the warnings on the label. Although it is this type of behaviour that limits which pesticides may be permitted for domestic use, it is still a concern that instructions are poorly understood or followed. Amateur application by householders is usually only for a brief period, typically less than 10 min, so exposure is usually confined to the finger using a pressure pack dispenser (Roff et al., 1998). Biological monitoring of pesticide exposures in residents living near agricultural land was implemented in 2011 to determine data collection procedures (Galea et al., 2011) prior to surveys planned in subsequent years.

(a)

(b)

Fig. 5.9 Examples of houses close to treated field, without a garden or hedge between house and sprayer (a) Norfolk (Photograph by Vincent Fallon); (b) sprayer near a house but fitted with an airsleeve to direct spray downward into the crop (Photograph by Alison Craig).

Apart from potential direct exposure when using a pesticide, residents can take residues of pesticides into their houses on clothing, especially agricultural workers, on shoes by walking over treated surfaces. In areas with fleas on animals, carpets and furniture, residents will also be exposed to insecticides by touching treated surfaces.

Residue transfer from a treated surface can be assessed using whole-body dosimeters (all-cotton suits, cotton socks and cotton gloves are worn). A standardised exercise routine (referred to as Jazzercize) is followed to represent daily human activities to assess exposure by maximum contact with treated surfaces (Krieger et al., 2001). These studies can be combined with bio-monitoring pesticides or their breakdown products detected in urine samples. In one study 1.6 mg of the insecticide chlorpyrifos was extracted from the whole-body dosimeters, which was significantly below the amount estimated using US EPA-operating procedures. Urine samples indicated only 1.3 μg chlorpyrifos had been absorbed (Bernard et al., 2001). Other studies have indicated that the human skin removes substantially less pesticide residue from treated surfaces than when a surface is wiped or a polyurethane foam roller is used to sample a surface. Hand contact averaged approximately 1–6 ng/cm^2 of chlorpyrifos-treated carpet contacted, that is less than 1% of the amount deposited on the carpet 3.5 h earlier (Lu and Fenske, 1999). Brouwer et al. (1999) found adherence of 1.07 $\mu g/cm^2$ after 12 contacts of a hand on a contaminated surface, a transfer efficiency of less than or equal to 2%.

Sampling in a metropolitan and an agricultural area of Washington State, USA, revealed quantifiable levels of azinphos methyl and chlorpyrifos on children's hands or their toys, that suggested a greater potential exposure in agricultural families (Lu et al., 2004). This contrasted with dietary studies as more samples of food for non-agricultural children had quantifiable amounts of OP pesticides. One application technique used in residences is referred to as a 'crack and crevice' treatment (Fig. 5.10).

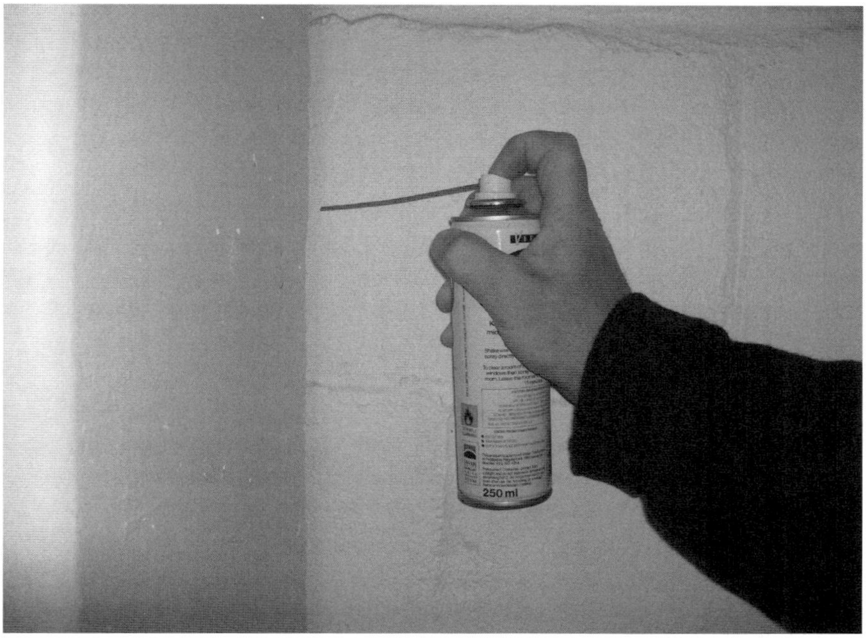

Fig. 5.10 A crack and crevice treatment with small applicator (Photograph by Graham Matthews).

The aim is to get insecticide into cracks in surfaces, such as brickwork, where cockroaches and other nuisance insects may hide. In one study in the USA 1.29 g of chlorpyrifos was applied to 'harbourages' in the kitchen using 259 ml of liquid (Stout and Mason, 2003). Over a 21-day study, chlorpyrifos was dispersed into the air and was distributed to other parts of the house, being detected on surgical sponge samplers. The deposition was highest in the kitchen, but decayed over the sampling period (Fig. 5.11). The data confirmed that crack and crevice treatments minimise human exposure compared to the use of pressure packs (total release aerosols) (Mason et al., 2000).

In an earlier study in California, house dust samples from 11 homes and hand-wipe samples from children in each house were analysed for 33 and 9 pesticides, respectively (Bradman et al., 1997). Ten pesticides, including chlorpyrifos and diazinon were detected with amounts greater in homes with a farm worker compared to those with no farm worker. Apart from two houses, levels detected were below 33 ppm. 'Take-home' residues of pesticides is clearly a significant factor in children's exposure to pesticides in rural areas (Garry, 2004), especially where the more toxic insecticides are used in agriculture, but uneducated use of dispensers in a home is also a crucial source of exposure to pesticides. Lu et al. (2000) also reported ×5 higher concentrations of pesticide metabolites in children living in agricultural areas compared to reference children with 0.01 μg/ml.

Epidemiological studies have been conducted to determine whether there is an association between the close proximity of residences to farmland and the incidence of illness. In one study in California, the incidence of breast

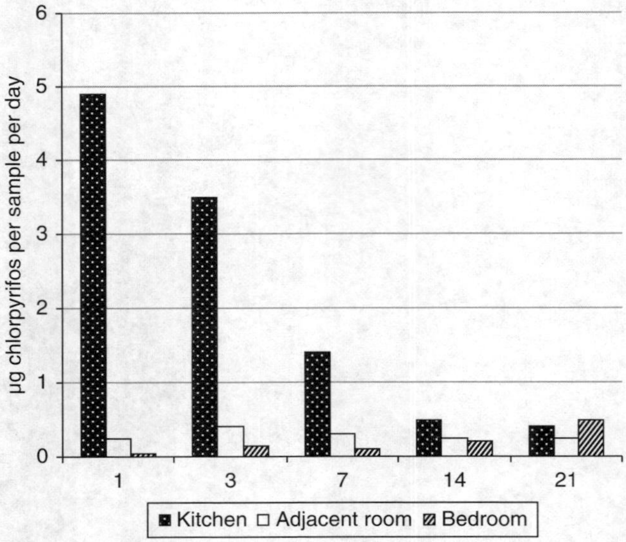

Fig. 5.11 Distribution of an insecticide after crack and crevice treatment in a house (Adapted from Stout and Mason, 2003).

cancer was studied among the 329,000 active and retired females enrolled in the State Teachers Retirement Scheme (Reynolds et al., 2004). Their analyses suggested that the incidence of breast cancer was not elevated in areas where there had been a recent high use of pesticides. In a study in two counties in the UK, spatial association between three selected pesticides, and breast cancer incidence rates was found in rural Leicestershire, but not in Lincolnshire or in urban areas (Muir et al., 2004). Three of the four pesticides were aldicarb (used as a granule for nematode control in potato fields), atrazine (herbicide) and lindane (used in buildings to treat timber) as these were considered to be 'oestrogenic'. The techniques were able to rule out strong associations, but a much larger and expensive study of individuals would be required to detect a smaller risk.

A meta-analysis on the effects of atrazine considered that due to poor-quality data and the lack of robust findings across studies, conclusions about a causal link between atrazine and adverse pregnancy outcomes are not warranted (Goodman et al., 2014).

In examining whether there was an association between childhood cancer and pesticide exposure, Gunier et al. (2001) examined the use of three pesticides with similar toxicologic properties, namely propargite, methyl bromide and trifluralin in California and concluded that pesticide use density would be of value in studying childhood cancer. Following this approach, Sarigiannis et al. (2013) developed an inventory of pesticide emissions in the atmosphere, which showed that atmospheric emission of pesticides varies significantly across Europe due to different usage patterns depending on crops being cultivated and agro-climatic conditions.

In Japan, studies of the exposure of young children in houses and childcare facilities were conducted following the aerial application of OP insecticides, including the volatile dichlorvos on rice using a remotely controlled helicopter. Inhalation exposure indoors at childcare facilities was comparable with, or more than, that at home and correlated inversely with the distance from the treated farm (Kawahara et al., 2005).

Levels of organochlorine pesticides, including DDT and its metabolites were detected in 200 women living in an area of Spain with intensive greenhouse agriculture who provided adipose tissue and blood samples during surgery. As these chemicals can be mobilised in the body during pregnancy and lactation, health consequences for the children of those exposed is of some concern and further research is needed on infant exposure in this way (Botella et al., 2004). Fortunately as the organochlorines have been banned, except DDT for vector control, levels of these pesticides in humans are expected to decline.

In a recent study in the USA, the potential association of congenital heart defects with residential proximity to commercial agricultural pesticide applications in the San Joaquin Valley, California, was examined, but for most pesticides there was no association with increased risk of specific heart defect phenotypes. The authors considered that in a few cases the results need to be interpreted with caution until replicated in other study populations (Carmichael et al., 2014).

When Li et al. (2014) analysed air samples in different seasons in an urban community in Guangzhou, China, chlorpyrifos and cypermethrin were the most dominant pesticides detected in the atmosphere, accounting for 68 and 15% of the total respectively. Potential exposure by inhalation suggested that children, toddlers and infants had the highest exposure, but the risk quotients were low for all age groups when annual average concentrations were used as exposure metrics. Exposure risk was higher in summer and autumn than the annual average level due to higher atmospheric pesticide concentrations, longer exposure times. Other studies have also indicated that children should not be exposed to pesticides (VanMaele-Fabry et al., 2011).

In some areas of the world, especially in tropical countries, residential areas are treated with insecticides to control mosquitoes and other vectors of disease. This can be in the form of a residual spray, a space treatment or by sleeping under a treated bed-net. Residual sprays on the walls of dwellings with DDT wettable powder at 2 g ai/m^2 was the standard technique with compression sprayers to break the transmission of malaria in the 1950s. DDT is still used in areas with major malaria problems and in South Africa had a distinct advantage over the use of a pyrethroid, although concerns about its use were raised. One concern was that some of the DDT would be sprayed on crops, on which its use is banned. In South Africa a pilot scheme has used a mobile phone app (mSpray) to determine the location of malaria cases in relation to the location of indoor residual spraying. The aim of mapping where the malaria was occurring was to improve the judicious use of insecticides (Eskenazi et al., 2014).

WHO has been promoting the use of treated bed-nets as a means of protecting young children from malaria. The nets can be treated by soaking the netting in a container with a suspension of a pyrethroid insecticide, but the net material is now impregnated with insecticide before the net is manufactured. The residual activity of a pyrethroid treated bed net is 5 years, but this means that selection for resistance now occurs throughout the year with some mosquitoes encountering such nets even in the dry season. Although washing removes the insecticide on the surface of the net, the insecticide migrates to the surface from within the fibre, so the net is effective again after a short delay. Risk assessments have shown that this technique is extremely safe for the occupants of houses. Other techniques such as the use of heated dispensers, mosquito coils and small aerosol dispensers are used by individual householders.

In some residential areas, insecticide is applied at a very low dose by fogging. Various types of thermal and cold foggers, either hand-carried or vehicle mounted may be deployed by health authorities. Crucially the fog is applied only when the insects are actively flying. This timing will depend on the target vector but is usually in the evening or early morning. In the USA, although malaria is not the problem, mosquito control has been a major factor in areas such as Florida, where extensive wet areas provide ideal breeding sites for mosquitoes. Apart from a risk of exposure to encephalitis, the need for mosquito control has been emphasised in recent years by the spread across the USA of West Nile virus. This virus first detected in New York, following the death of many crows has spread to humans from birds by mosquito vectors. Despite the need to stop the disease spread, extensive use of space treatments with insecticide has caused concern. A survey in New York City showed that exposure estimates of pyrethroids and dimethyl organophosphates were higher than in the United States overall, indicating the importance of considering pest and pesticide burdens in cities, when formulating pesticide use regulations (McKelvey et al., 2013).

In a study (Knepper et al., 2003) in a park a 20% permethrin plus 20% PBO formulation diluted 1:2 with water was applied at 0.0021 kg ai/ha as a cold fog. Filter papers placed on different surfaces were subsequently analysed to show after 12 h post-treatment a deposit of 0.66 ng/cm^2. The WHO ADI of permethrin is 0.05 mg/kg/day for the lifetime of an individual. If a child is playing in the park with a ball 28.2 cm diameter, that is with a surface area of approximately 2500 cm^2 and half the area picks up the permethrin residue from the surfaces on which it bounces, then the ball could pick up 827 ng. Assuming all of this is transferred to the surface of a child's hand and 10% is absorbed through the skin then the total amount of permethrin entering the child's body is calculated to be 82.7 ng. If the child weighed 25 kg, the exposure equals 3.3 ng/kg, which is over 15 000 times less than the ADI.

Thus the very small dose used in these space treatments is not expected to cause any harm to children and less so to adults. In another study in the USA, comparing children of farm workers with other children living in the same agricultural community, exposure to pesticides was affected by several factors, depending on how much time was spent outdoors and whether the children accompanied their parents to work in the fields (Shepherd-Banigan et al., 2014). In the developing countries, pesticides may be stored in houses, including the bedrooms as this is considered one of the safest places to avoid theft (Fig. 5.12). The pesticides are sometimes purchased and stored in unlabelled containers (Ngowi et al., 2001a; Matthews et al., 2003). Unfortunately medical services in rural areas do not have the required training in toxicity of pesticides and treatment in cases of poisoning (Ngowi et al., 2001b).

Analysis for persistent insecticides in blood samples of 155 volunteers from 13 different areas of the UK showed DDT was still present although its use had been discontinued for many years. Blood samples also revealed other compounds including PCBs (Table 5.3) (Anon, 2003).

Fig. 5.12 Pesticide container in the bedroom of a house of an African villager (Photograph by Graham Matthews).

Table 5.3 Results of blood samples in UK in 2003 (amounts in ng/g lipid)

Chemical	Minimum	Maximum	Median
Total chemical	46	2024	360
Total DDT and metabolites	1.3	1715	107
Total organochlorine	7.1	1754	140
Total PCB*	14	665	169
No. of chemicals detected/78	9	49	27

1 ng = 0.000000001 g. While this quantity is extremely minute, it is at this nano level that reactions occur at cellular level. However, DDT is known to be stored in fatty tissues and has not been shown to affect man adversely when exposed to large amounts of it.
* Brominated flame retardants are among the PCBs to which we are exposed.

Worker exposure

Many people come into contact with surfaces treated with pesticides as part of their normal working day. These include those harvesting many vegetable and fruit crops and flowers, especially on protected cropping, where deposits may remain on foliage longer than when crops are exposed to rain. Care is particularly needed to avoid entry into and touching treated crops immediately after a spray application. Hands and other parts of the body touching treated surfaces may be wetted by spray deposits or when the deposit has dried, they may pick up dry, dislodgeable residues later (Van Hemmen and Brouwer, 1997). The amount of dislodgeable residues is normally related to the dosage applied, but will be affected by the formulation used and affinity of particles to the foliar surfaces. Their distribution will also be a function of the application technique. Exposure to these deposits is predominantly dermal and can also be intermittent, thus from a toxicological point of view, Hakkert (2001) suggested setting more than one Acceptable Operator Exposure Level (AEOL) covering different periods of exposure.

As part of the BROWSE project, the considerable number of worker exposure scenarios (Ngoc et al., 2014) was confined to four main parts:

1. Dermal exposure outdoors related to harvesting, pruning and thinning orchard crops, soft fruit and vineyards
2. Dermal exposure indoors related to harvesting ornamentals and fruiting vegetables
3. Exposure to vapour in open fields and protected cropping.
4. And sowing treated seeds.

The dermal exposure model is based on the algorithm used in EUROPOEM and by the U.S. EPA

$$\mathrm{DE} = \mathrm{TC} \times \mathrm{DFR} \times T$$

where DE = dermal exposure, TC = transfer coefficient, DFR = dislodgeable foliar residue and T is the duration of the activity.

Assessments of exposure of workers harvesting or otherwise touching treated plants is often by means of hand washing or skin wipes, using both chemical and mechanical action to transfer surface deposits to the sampling surface (Brouwer et al., 2000a).

The methods are relatively inexpensive, but wipe sampling does not remove as much compared to hand wash samples. Sampling is also influenced by the period between exposure and sampling, as the sample represents only what is still accessible to the sampling technique when this is done.

In a study in greenhouses in which chrysanthemums were being grown, dermal and inhalation exposure was measured during high volume spray application of the OP insecticide methomyl. Subsequent studies using the fungicide chlorothalonil continued the study for re-entry and harvesters (Brouwer et al., 1994). The data provided a useful comparison of the exposure on hands during mixing the pesticide, spraying and either manually or automatic harvesting, which occurred on average 27 days after treatment (Table 5.4) to use in risk management. Inhalation exposure was generally less than 0.01 mg/h. Wearing gloves was shown to significantly lower actual exposure (Table 5.5) (Brouwer et al., 2000a) Other studies in Holland (Brouwer et al., 1992) have examined cold fogging (called low-volume misting) in greenhouses. With the volume median diameter of droplets less than 50 µm, the threshold limit value of

Table 5.4 Mean potential dermal exposure of hands (mg/h) (Brouwer et al., 1994)

Mixing and loading sprayer	13.0
During high volume application	0.8
Harvesting	
Manual	3.6
Automatic	1.1

Table 5.5 Actual exposure of hands in μg the insecticide propoxur* expressed as median value and range (Brouwer et al., 2000c)

Operation	Gloves	No gloves
Mixing/loading	8 (7–148)	231 (8–5785)
Application	8 (4–47)	122 (9–416)
Total spraying	16 (11–56)	348 (31–2390)
Harvesters	8 (5–299)	164 (7–1523)

*Applied at 36.8 ± 14.3 g ai/1000 m^2 at an average volume rate of 113.5 ± 44.5 l/1000 m^2 over an average period of 36 min.

a volatile insecticide was still substantially exceeded 6 h after application, but venting the greenhouse for 1 h then permitted safe entry. Other glasshouse studies in California showed that a fog slowly settled during the night and re-admission was possible after venting (Giles et al., 1995). The cold fogging resulted in a more homogeneous distribution within the crop compared to high volume spraying (Brouwer et al., 2000b). Comparing the impact of chemical pest management with an integrated pest management programme in two experimental greenhouses Anton et al. (2004) considered that the selection of chemicals could be more important than choice of control programme in relation to life cycle assessment of pesticides, including human exposure routes.

If the average air concentration of an insecticide downwind of a crop being sprayed is 30 μg/m^3, over a 5 h period, then reducing to 5 μg/m^3, the time weighted average (TWA) is 10.2 μg/m^3, so assuming a breathing rate of 15 m^3/day, 100% absorption and a body weight of 60 kg the exposure is 0.15 mg/person or 0.00255 mg/kg bw/day. This value is then compared with the AOEL and ADI of the pesticide being considered. For a smaller person weighing only 25 kg, the exposure would be correspondingly higher 0.006 mg/kg bw/day, at the same breathing rate. Remoundou et al. (2014) have reviewed pesticide exposure and provide a comprehensive overview aimed at assisting policy-makers, and risk communicators in the development of targeted training and awareness-raising material for operators, workers, bystanders and residents.

In risk assessment for those who may have to enter crops and more generally in relation to the environment, registration authorities would like to have some indication of the amount of pesticide intercepted by a crop and the period over which deposits are retained. There has been an attempt to produce standardised values for foliar interception (Linders et al., 2000), but the results will vary significantly between methods of application, the crop and subsequent weather conditions.

This chapter has provided information on the movement of spray from equipment and how bystanders, and those living or working in areas where pesticides are used may be exposed to the chemicals. Careful choice of pesticides, avoidance of highly volatile chemicals and changes in spray techniques, including the use of coarser sprays, shields and directed air assist sprayers – drift reduction technology (DRT) – have reduced the exposure to those who are not directly involved in the application processes. In some countries, buffer zones (see Chapter 6) are already required to protect sensitive areas, such as housing, schools and hospitals. With the reform of the Common Agricultural Policy, farmers seeking the Single Farm Payment must 'set aside' part of their arable land. The minimum width of land accepted as set-aside has been reduced from 10 to 5 m, so there should not be any impediment to these areas being buffer strips provided they are managed without spraying pesticides. Spot herbicide treatments to control certain weeds may still be required.

References

Anon (2003) National Biomonitoring Survey, 2003. WWF, UK.

Anon (2004) Pesticide Incidents Report 1 April 2003–31 March 2004. HSE, Bootle.

Anton, A., Castells, F., Montero, J.I. and Huijbregts, M. (2004) Comparison of toxicological impacts of integrated and chemical pest management in Mediterranean greenhouses. *Chemosphere* **54**, 1225–1235.

Bernard, C.E., Nuygen, H., Truong, D. and Krieger, R.I. (2001) Environmental residues and biomonitoring estimates of human insecticide exposure from treated residential turf. *Archives of Environmental Contamination and Toxicology* **41**, 237–240.

Botella, B., Crespo, J., Rivas, A., Cerrillo, I., Olea-Serrano, M.F. and Olea, N. (2004) Exposure of women to organochlorine pesticides in Southern Spain. *Environmental Research* **96**, 34–40.

Bradman, M.A., Harnly, M.E., Draper, W., Seidel, S., Teran, S., Wakeham, D. and Neutra, R. (1997) Pesticide exposures to children from California's Central Valley: results of a pilot study. *Journal of Exposure Analysis and Environmental Epidemiology* **7**, 217–234.

Briand, O., Bertrand, F., Seux, R. and Millet, M. (2002) Comparison of different sampling techniques for the evaluation of pesticide spray drift in apple orchards. *The Science of the Total Environment* **288**, 199–213.

Brouwer, D.H., Boeniger, M.F. and Van Hemmen, J. (2000a) Hand wash and manual skin wipes. *Annals Occupational Health* **44**, 501–510.

Brouwer, D.H., de Haan, M. and van Hemmen, J.J. (2000b) Modelling re-entry exposure estimates. Application techniques and rates In *Worker Exposure to Agrochemicals*. Honeycutt, R.C. and Day, Jr. E.W. (Eds.). ACS Symposium Series. CRC Lewis Publishers, Baton Rouge, FL, pp. 119–138.

Brouwer, D.H., de Vreede, J.A.F., Ravensberg, J.C. Engel, R. and van Hemmen, J.J. (1992) Dissipation of aerosols from greenhouse air after application of pesticides using a low volume technique. Implications for safe re-entry. *Chemosphere* **24**, 1157–1169.

Brouwer, D.H., de Vreede, J.A.F., de Haan, M, van der Vijver, L., Veerman, M., Stevenson, H. and van Hemmen, J.J. (1994) Exposure to pesticides during and after application in the cultivation of chrysanthemums in greenhouses. Health risk and risk management. Mededelingen–Faculteit Landbouwkundige En Toegepaste Biologische Wetenschappen, **59**, 1393.

Brouwer, D.H., de Vreede, S.A.F., Meuling W.J.A. and van Hemmen, J.J. (2000c) Determination of the efficiency for pesticide exposure reduction with protective clothing: field study using biological monitoring. In *Worker Exposure to Agrochemicals*. Honeycutt, R.C. and Day, Jr. E.W. (Eds.). ACS Symposium Series. CRC Lewis Publishers, Baton Rouge, FL, pp. 63–84.

Brouwer, D.H. Kroese, R. and van Hemmen, J.J. (1999) Transfer of contaminants from surface to hands: experimental assessment of linearity of the exposure process, adherence to the skin and area exposed during fixed pressure and repeated contact with surfaces contaminated with powder. *Applied Occupational and Environmental Hygiene* **14**, 231–239.

Butler Ellis, M.C. and Miller, P.C.H (2010) The Silsoe spray drift model: a model of spray drift for the assessment of non-target exposures to pesticides. *Biosystems Engineering*, **107**, 169–177.

Butler Ellis, M.C., Lane, A.G., O'Sullivan, C.M., Miller, P.C.H. and Glass, C.R. (2010a) Bystander exposure to pesticide spray drift: new data for model development and validation. *Biosystems Engineering*, **107**, 162–168.

Butler Ellis, M.C., Underwood, B., Peirce, M.J. Walker, C.T. and Miller, P.C.H. (2010b) Modelling the dispersion of volatilised pesticides in air after application for the assessment of resident and bystander exposure. *Biosystems Engineering*, **107**, 149–154.

Butler Ellis, M.C., Lane, A.G., O'Sullivan, C.M.,Alanis, R., Harris, A., Stallinga, H. and Van de Zande, J.C. (2014) Bystander and resident exposure to spray drift from orchard applications: field measurements, including a comparison of spray drift collectors. *Annals of Applied Biology*, **122**, 187–194.

Carmichael, S.I.,Yang, W.,Roberts, E., Kegley, S.E.,Padula, A.M.,English, P.B., Lammer, E.J., Shaw, G.M. (2014) Residential agricultural pesticide exposures and risk of selected congenital heart defects among offspring in the San Joaquin Valley of California. *Environmental Research* **135**, 133–138.

Cross, J.V., Walklate, P.J., Murray, R.A. and Richardson, G.M. (2001) Spray deposits and losses in different sized apple trees from an axial fan orchard sprayer. 1. Effects of spray liquid flow rate. *Crop Protection* **20**, 13–30.

Dobson, C.M., Minski, M.J. and Matthews, G.A. (1983) Neutron activation analysis using dysprosium as a tracer to measure spray drift. *Crop Protection* **2**, 345–352.

Douzals, J.P. and Heidary, M.A. (2014) How spray characteristics and orientation may influence spray drift in a wind tunnel. *Aspects of Applied Biology*, **122**, 271–278.

Driver J.H. and Whitmyre, G.K. (1996) Assessment of residential exposures to pesticides and other chemicals. Pesticide Outlook, 6–10.

Elliott J.G. and Wilson, B.J. (1983) The influence of weather on the efficiency and safety of pesticide application: the drift of herbicides. BCPC Occasional Publication 3, Farnham, UK.

Epple, J., Maguhn, J. Spitzauer, P. and Kettrup, A. (2002) Input of pesticides by atmospheric deposition. *Geoderma* **105**, 327–349.

Eskenazi, B. Quiros-Alcala, L. Lipsitt, J.M. Wu, L.D. Kruger, P. Ntimbane, T. Bruns Dawn, J. Bornman, M.S.R., and Seto E. (2014) mSpray: a mobile phone technology to improve malaria control efforts and monitor human exposure to malaria control pesticides in Limpopo, South Africa. *Environment International* **68**, 219–226.

EU (2008) Pesticides in Air: Considerations for Exposure Assessment. SANCO/10553/2006. European Union, Brussels.
Foque, D., Dekeyser, D., Zwertvaegher, I. and Nuyttens, D. (2014) Accuracy of multiple mineral tracer methodology for measuring spray deposition. *Annals of Applied Biology* **122**, 203–211.
Galea, K.S., MacCalman, L., Jones, K., Cocker, J., Teedon, P., Sleeuwenhoek, A.J., Cheerie, J.W. and van Tongeren, M. (2011) Biological monitoring of pesticide exposures in residents living near agricultural land. *BMC Public Health* **11**, 856.
Garry, V.F. (2004) Pesticides and children. *Toxicology and Applied Pharmacology* **198**, 152–163.
Gil, Y. and Sinfort, C. (2005) Emission of pesticides to the air during sprayer application: a bibliographic review. *Atmospheric Environment* **39**, 5183–5193.
Gilbert, A.J. and Bell, G.J. (1988) Evaluation of the drift hazards arising from pesticide spray application. *Aspects of Applied Biology* **17**, 363–376.
Giles, D.K., Welsh, A., Steinke, W.E. and Saiz, S.G. (1995) Pesticide inhalation exposure, air concentration and droplet size spectra from greenhouse fogging. *Transactions of the ASAE* **38**, 1321–1326.
Goodman, M., Mandel, J.S., DeSesso, J.M. and Scialli, A.R. (2014) Atrazine and pregnancy outcomes: a systematic review of epidemiologic evidence. *Birth Defects Research B* **101**, 215–236.
Grey, C.N.B., Nieuwenhuijsen, M.J., Golding, J. and the ALSPAC Team. (2004) The use and disposal of household pesticides. *Environmental Research* **97**, 109–115.
Grover, R. Maybank, J. and Yoshida, K. (1972) Droplet and vapor drift from butyl ester and dimethylamine salt of 2,4-D. *Weed Science* **20**, 320–324.
Gunier, R.B., Harnly, M.E., Reynolds, P., Hertz, A. and Von Behren, J. (2001) Agricultural pesticide use in California: pesticide prioritization, use densities, and population distributions for a childhood cancer study. *Environmental Health Perspectives* **109**, 1071–1078.
Hakkert, B.C. (2001) Refinement of risk assessment of dermally and intermittently exposed pesticide workers: a critique. *The Annals of Occupational Health* **45**, S23–S28.
Harrad, S. and Mao, H. (2004) Atmospheric PCBs and organochlorine pesticides in Birmingham, UK: concentrations, sources, temporal and seasonal trends. *Atmospheric Environment* **38**, 1437–1445.
Hatfield, J.L. Wesley, C.K. Prueger, J.H. and Pfeiffer, R.L. (1996) Herbicide and nitrate distribution in Central Iowa rainfall. *Journal of Environmental Quality* **25**, 259–264.
Himel, C.M. (1974) Analytical methodology in ULV. *British Crop Protection Monograph* **11**, 112–119.
Jensen P.K. and Olesen, M.H. (2014) Spray mass balance in pesticide application: a review. *Crop Protection* **61**, 23–31.
Johnstone, D.R. and Matthews, G.A. (1965) Evaluation of swath pattern and droplet size provided by aboom and nozzle installation fitted to a Hiller UH-12 helicopter. *Agricultural Aviation*, **7**(2), 46–50.
Kawahara, J., Horikoshi, R., Yamaguchi, T., Kumagai, K. and Yaragisawa, Y. (2005) Air pollution and young children's inhalation exposure to organophosphorus pesticide in an agricultural community in Japan. *Environmental International* **31**, 1123–1132.
Kennedy, M.C., Butler-Ellis, M.C. and Miller, P.C.H. (2012) BREAM: a probabilistic bystander and resident exposure assessment model of spray drift from an agricultural boom sprayer. *Computers and Electronics in Agriculture* **88**, 63–71.
Knepper, R.G. Walker, E.D. and Kamrin, M.A. (2003) ULV studies of permethrin in Saginaw, Michigan. *Wing Beats* **14**, 22–23, 32–33.

Krieger, R.I. Dinoff, T.M. and Ross, J.H. (2001) Pesticide exposure assessment: Jazzercize™ activities to determine extreme case indoor exposure potential and in-use biomonitoring. In Honeycutt, R.C. and Day, Jr. E.W. (Eds.). *Worker Exposure to Agrochemicals*. CRC Lewis Publishers, Boca Raton, FL, pp. 97–106.

Lee, S., McLaughlin, R., Harnly, M., Gunier, R. and Kreutzer, R. (2002) Community exposures to airborne agricultural pesticides in California: ranking of inhalation risks. *Environmental Health Perspectives* **110**, 1175–1184.

Li, H., Mad, H., Lydy, M.J. and You, J. (2014) Occurrence, seasonal variation and inhalation exposure of atmospheric organophosphate and pyrethroid pesticides in an urban community in South China. *Chemosphere* **95**(2014), 363–369.

Linders, J., Mensink, H., Stephenson, G. Wauchope, D. and Racke, K. (2000) Foliar interception and retention values after pesticide application. A proposal for standardized values for environmental risk assessment. *Pure and Applied Chemistry* **72**, 2199–2218.

Lloyd, G.A. and Bell, G.J. (1983) Hydraulic Nozzles: Comparative Drift Study. MAFF Report SC 7704, London, UK.

Lloyd, G.A., Bell, G.J., Samuels, S.W., Cross J.V. and Berrie, A.M. (1987) Orchard Sprayers: Comparative Operator Exposure and Spray Drift Study. MAFF Report, London, UK.

Lu C. and Fenske, R.A. (1999) Dermal transfer of chlorpyrifos residues from residential surfaces: comparison of hand press, hand drag, wipe and polyurethane foam roller measurements after broadcast and aerosol pesticide applications. *Environmental Health Perspectives* **107**, 463–467.

Lu, C. Fenske, R.A., Simcox, N.J. and Kalman, D. (2000) Pesticide exposure of children in an agricultural community: evidence of household proximity to farmland and take home exposure pathways. *Environmental Research Section* **A84**, 290–302.

Lu, C., Kedan, G. Fisker-Andersen, J., Kissel, J.C. and Fenske, R.A. (2004) Multipathway organophosphorus pesticide exposures of pre-school children living in agricultural and non-agricultural communities. *Environmental Research* **96**, 283–289.

Mason, M.A., Sheldon, I.S. and Stout II, D.M. (2000) The Distribution of Chlorpyrifos in Air, Carpeting and Dust and its Reemission from Carpeting Following the Use of Total Release Aerosols in an Indoor Air Quality Test House. *Proceedings of the Symposium on Engineering Solutions to Indoor Air Quality Problems*. Raleigh, NC, 17–19 July, pp. 92–102.

Mathers, J.J., Glass, C.R., Harrington, P. and Smith, S. (2006) Techniques for estimation of bystander exposure. *Aspects of Applied Biology* **77**, 253–257.

Matthews, G.A. and Johnstone, D.R. (1968) Aircraft and tractor spray deposits on irrigated cotton. *Cotton Growing Review* **45**, 207–218.

Matthews, G.A., Wiles, T. and Baleguel, P. (2003) A survey of pesticide application in Cameroon. *Crop Protection* **22**, 707–714.

McKelvey, W., Jacobson, J.B., Kass, D., Barr, D.B., Davis, M., Calafat, A.M. and Aldous, K.M. (2013) Population-based biomonitoring of exposure to organophosphate and pyrethroid pesticides in New York city. *Environmental Health Perspectives* **121**, 1349–1356.

Miller, D.R. and Stoughton, T.E. (2000) Response of spray drift from aerial applications at a forest edge to atmospheric stability *Agricultural and Forest Meteorology* **100**, 49–58.

Miller, P.C.H. (1993) Spray drift and its measurement. In Matthews, G.A. and Hislop, E.C. (Eds.). *Application Technology for Crop Protection*. CABI, Wallingford, pp. 101–122.

Miller, P.C.H. and Hadfield, D.J. (1989) A simulation of the spray drift from hydraulic nozzles. *Journal of Agricultural Research* **42**, 135–147.

Miller, P.C.H., Hislop, E.C., Parkin, C.S., Matthews, G.A. and Gilbert, A.J. (1993) The Classification of Spray Generator Performance Based on Wind Tunnel Assessments of Spray Drift. *ANPP-British Crop Protection Council Second Symposium on Pesticide Application Techniques*, Strasbourg, pp. 109–116. ANPP, Paris.

Muir, K., Rattanamongkolgl, S., Smallman-Raynor, M., Thomas, M., Downer, S. and Jenkinson, C. (2004) Breast cancer incidence and its possible spatial association with pesticide application in two counties of England. *Public Health* **118**, 513–520.

Neumann, M. Schulz, R. Schafer, K., Muller, W. Mannheller, W. and Liess, M. (2002) The significance of entry routes as point and non-point sources of pesticides in small streams. *Water Research* **36**, 835–842.

Ngoc, K.D., Spanoghe, P., Van den Berg, E., Charistou, A., Glass, C.R., Frewer, L., Gerritsen-Ebben, R., Butler-Ellis, C., Capri, E. and Hart, A. (2014) Worker re-entry within the framework of BROWSE project. *Annals of Applied Biology*, **122**, 159–164.

Ngowi, A.V.F., Maeda, D.N., Wesseling, C., Partenen, T.J. Sanga, M.P. and Mbise, G. (2001a) Pesticide-handling practices in agriculture in Tanzania: observational data from 27 coffee and cotton farms. *International Journal of Occupational and Environmental Health* **7**, 326–332.

Ngowi, A.V.F., Maeda, D.N. and Partenen, T.J. (2001b) Assessment of the ability of health care providers to treat and prevent adverse health effects of pesticides in agricultural areas of Tanzania. *International Journal of Occupational Medicine and Environmental Health* **14**, 349–356.

Nuyttens, D., Taylor, W.A., de Schampheleire, M., Verboven, P. and Dekeyser, D. (2009) Influence of nozzle type and size on drift potential by means of different wind tunnel evaluation methods. *Biosystems Engineering* **103**, 271–280.

Piggott S.J. and Matthews G A (1999) Air induction nozzles: a solution to spray drift? *International Pest Control* **41**, 24–28.

Ramaprasad, J., Tsai, M-Y., Elgetthun K., Hebert, V.R. Felsot, A., Yost, M.G. and Fenske, R.A. (2004) The Washington aerial spray drift study: assessment of off-target organophosphorus insecticide atmospheric movement by plant surface volatilization. *Atmospheric Environment* **38**, 5703–5713.

RCEP (2005) *Crop Spraying and the Health of Residents and Bystanders*. Royal Commission on Environmental Pollution, London.

Remoundou, K., Brennan, M., Hart, A. and Frewer, L.J. (2014) Pesticide risk perceptions, knowledge, and attitudes of operators, workers, and residents: a review of the literature. *Human and Ecological Risk Assessment*, **20**, 1113–1138.

Reynolds, P., Hurley, S.E. Goldberg, D.E., Yerabati, S. Gunier, R.B., Hertz, A., Anton-Culver, H., Berstein, L., Deapen, D., Horn-Ross, P.L., Peel, D., Pinder, R., Ross, R.K. West, D., Wright, W.E. and Ziogas, A. (2004) Residential proximity to agricultural pesticide use and incidence of breast cancer in the California teachers study cohort. *Environmental Research* **96**, 206–218.

Rimmer, D.A., Johnson, P.D., Kelsey, A., Warren, N.D. and Saunders, C.J. (2006) Spray trials to assess approaches for post-event incident investigation. *Annals of Applied Biology* **77**, 259–266.

Roff, M., Baldwin, P., Thompson, J. and Wheeler, J. (1998) Consumer exposure arising from the application of indoor pesticides. *Journal of Aerosol Science* **29**, S1291–S1292.

Sacchettini, G., Calliera, M., Marchis, A., Lamastra, L. and Capri, E. (2012) The stakeholder-consultation process in developing training and awareness-raising material within the framework of the EU Directive on Sustainable Use of Pesticides: the case of the EU-project BROWSE. *Science of the Total Environment* **438**, 278–285.

Sarigiannis, D.A., Kontoroupis, P., Solomou, E.S., Nikolaki, S., and Karabelas, A.J. (2013) Inventory of pesticide emissions into the air in Europe. *Atmospheric Environment* **75**, 6–14.

Scheyer, A., Graeff, C., Morville, S., Mirabel, P. and Millet, M. (2005) Analysis of some organochlorine pesticides in an urban atmosphere (Strasbourg, east of France). *Chemosphere* **58**, 1517–1524.

Schulz, R. (2001) Comparison of spray drift and run-off-related input of azinphos-methyl and endosulfan from fruit orchards into the Lourens River, South Africa. *Chemosphere* **45**, 397–407.

Shepherd-Banigan, M., Ulrich, A. and Thompson, B. (2014) Macro-activity patterns of farmworker and non-farmworker children living in an agricultural community. *Environmental Research* **132**, 176–181.

Siebers, J. Binner, R. and Wittich K-P. (2003) Investigation on downwind short-range transport of pesticides after application in agricultural crops. *Chemosphere* **51** 397–407.

Stoughton, T.E., Miller, D.R. Yang, X. and Ducharme K.M. (1997) A comparison of spray drift predictions to lidar data. *Agricultural and Forest Meteorology* **88**, 15–26.

Stout II, D.M. and Mason M.A. (2003) The distribution of chlorpyrifos following a crack and crevice type application in the US EPA Indoor Air Quality Research House. *Atmospheric Environment* **37**, 5539–5549.

Teil, M-J., Blanchard, M. and Chevreuil, M. (2004) Atmospheric deposition of organochlorines (PCBs and pesticides) in northern France. *Chemosphere* **55**, 501–504.

Tsai, M-Y., Elgethun, K., Ramaprasad, J., Yost, M.G., Felsot, A.S., Hebert, V.R. and Fenske, R.A. (2005) The Washington aerial spray drift study: modeling pesticide spray drift deposition from an aerial application. *Atmospheric Environment* **39**, 6194–6203.

Turnbull, A.B. (1995) An Assessment of the Fate and Behaviour of Selected Pesticides in Rural England. PhD thesis. University of Birmingham, Birmingham.

Unsworth, J.B., Wauchope, R.D., Klein, A.W., Dorn, E., Zeeh, B., Yeh, S.M. Akerblom, M., Racke, K.D. and Rubin, B. (1999) Significance of the long range transport of pesticides in the atmosphere. *Pure and Applied Chemistry*, **71**, 1359–1383.

Van de Zande, J.C., Butler-Ellis, M.C., Wenneker, M., Walklate, P.J. and Kennedy, M. (2014) Spray drift and bystander risk from fruit crop spraying. *Annals of Applied Biology* **122**, 177–185.

Van Dijk, H.F.G. and Guicherit, R. (1999) Atmospheric dispersion of current-use pesticides – a review of the evidence from monitoring studies. *Water, Air, and Soil Pollution* **115**, 21–70.

Van Hemmen, J.J. and Brouwer, D.H. (1997) Exposure assessment of pesticides: operators and harvesters risk evaluation and risk management. *Mededelingen – Faculteit Landbouwkundige En Toegepaste Biologische Wetenschappen* **62/2** 113–130.

Van Maele-Fabry, G., Lantin, A-C., Hoet, P. and Lison, D. (2011) Residential exposure to pesticides and childhood leukaemia: a systematic review and meta-analysis. *Environment International* **37**, 280–291.

Walls, J., Rowe, G. and Frewer, L. (2010) Stakeholder engagement in food risk management: evaluation of an iterated workshop approach. *Public Understanding of Science* **20**, 241–260.

Warren, L.M. (2009) Analysis – Healthy crops or healthy people? Balancing the needs for pest control against the effect of pesticides on bystanders: Secretary of State for Environment Food and Rural Affairs v Georgina Downs (Court of Appeal (Civil Division)) [2009] EWCA Civ 664. *Journal of Environmental Law* **21**, 483–499.

Wenneker, M., Van de Zande, J.C., Stallinga, H., Michielsen, J.M.P.G., Van Velde, P. and Nieuwenhuizen, A.T. (2014) Emission reduction in orchards by improved spray deposition and increased spray drift reduction of multiple row sprayers. *Annals of Applied Biology* **122**, 195–202.

Fig. 1.9 Cocoa farmers need apply fungicides to protect pods from black pod disease (Photograph by Roy Bateman).

(a)

(b)

Fig. 2.2 Different containers. (a) Sachets (Photograph courtesy of Syngenta). (b) Container with built-in measure (Photograph by Graham Matthews).

(a)

(b)

Fig. 2.5 (a) FasTrans closed transfer system on container. (b) FasTrans system showing connection to sprayer (Wisdom Systems – eziserv Ltd.).

(f)

Fig. 3.2 Tractor sprayer (f) vertical boom sprayer in a glasshouse (Photograph by Richard Glass).

(c)

Fig. 3.4 Air-assisted sprayers for bush and tree crops. (c) three-row Munckhof air system (Photograph by Jerry Cross).

(d)

Fig. 3.5 Aerial spraying (d) UAV spraying a crop (Photograph by K. Giles, UC Davis, USA).

(a)

Fig. 3.10 (a) Nozzles protected by a shield to exclude downwind drift of spray for precision application of herbicides Varidome (Photograph courtesy of Micron).

(c)

Fig. 3.11 Rotary atomiser CDA spinning disc sprayer (c) 'Electrafan' sprayer being used in a glasshouse (Photograph courtesy of Micron).

(c)

Fig. 3.12 (c) Compression sprayer with electric pump (Photograph courtesy of Gloria).

(d)

Fig. 3.13 (d) Knapsack with electric pump to eliminate manual pumping (Photograph courtesy of Birchmeier).

(d)

Fig. 3.16 Space treatments: (d) vehicle mounted cold fogger (Photograph courtesy of Micron).

(c)

Fig. 4.3 (c) holding lance downwind of body to minimise operator exposure to spray (Photograph by Graham Matthews).

(b)

Fig. 6.14 (b) Wild flowers in field border (Photograph by Graham Matthews).

6 Environmental aspects of spray drift

In addition to the concerns about pesticides affecting the human population directly, registration authorities are also concerned with more general effects in the environment on other non-target organisms, which can also have indirect effects on people.

Protecting water

A major consideration in protecting the environment from exposure to pesticides has been to minimise spray droplets drifting and subsequently sedimenting on water surfaces. In Europe the Water Framework Directive requires all surface waters to meet good ecological status, that is a healthy ecosystem and under the Sustainable Use of Pesticides Directive EU Member States have to develop National Action Plans with objectives, targets and measures to reduce the risks associated with applying pesticides. Measures relating to changing the landscape structure include using buffer strips of plants alongside rivers, or by creating retention ponds and ditches to reduce pesticide exposure in surface waters.

Many of the studies on spray drift, referred to in the previous chapter have concentrated on the amount of pesticide collected on flat sampling surfaces within a relatively short distance downwind, rather than measurements of airborne drift. Studies in Germany (Ganzelmeier et al., 1995; Ganzelmeier and Rautmann, 2000) and others (e.g. Hewitt, 2000) have provided data to support legislation requiring no-spray or 'buffer' zones, the width of which depends on the type of pesticide and risk assessments in relation to fish and other aquatic organisms (Fig. 6.1). It also takes account of the need to minimise exposure so that water extracted for drinking meets the EU standards, namely 0.1 μg/l for a single pesticide and 0.5 μg/l for all pesticides. There is also a standard of 0.03 μg/l for certain pesticides (e.g. organochlorine).

Pesticides: Health, Safety and the Environment, Second Edition. G.A. Matthews.

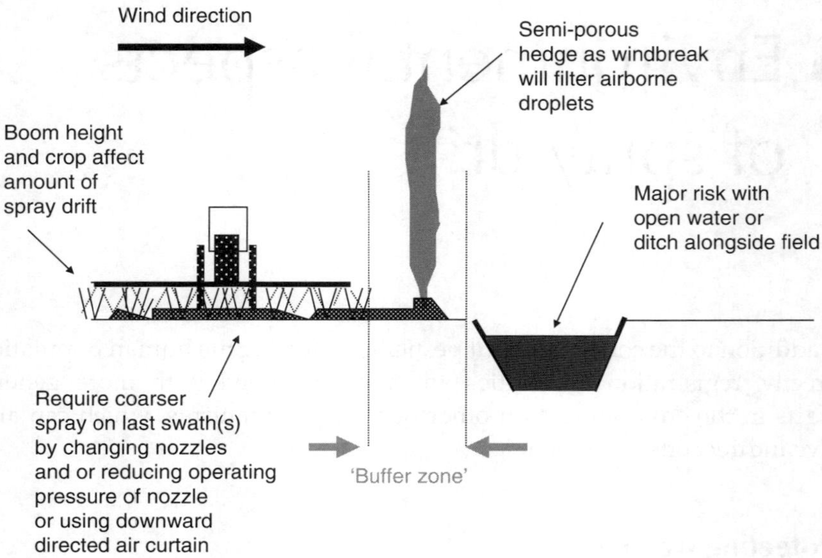

Fig. 6.1 Diagram showing position of a buffer zone to protect a watercourse.

The effectiveness of grass-covered buffer strips at the downwind and lower edges of fields will vary depending on a number of factors apart from its width (Reichenberger et al., 2007). Vegetation has been demonstrated to filter out spray drift, but surface run-off can also occur depending on the extent to which rain infiltrates the soil, the amount of rainfall and area that slopes towards a ditch. Modelling run-off Cardinali et al. (2013) considered that most herbicide loss (98%) is when an extreme rainfall event occurs possibly every 25 years in contrast to ordinary rainfall events.

Arias-Estevez et al. (2008) reviewed information on the influence of the physical and chemical characteristics of a soil system, such as moisture content, organic matter and clay contents, and pH, on the sorption/desorption and degradation of pesticides and their access to groundwater and surface waters. In the USA, a survey of over 1000 wells distributed nationally was carried out once between 1993 and 2001 and again between 2001 and 2011 to determine whether the occurrence of 83 pesticides could be detected in the water. The concentration of pesticides in 36 out of 58 well networks changed significantly between decades, mostly related to the most frequently detected pesticides (herbicides), but the changes were very small, ranging from 0.09 to 0.03 μg/l and well below human health benchmarks (Toccalino et al., 2014).

In France, Carluer et al. (2011) concluded that none of the buffer solutions examined removed pesticide 100% due to variations in hydrological flow (surface run-off, deep infiltration, lateral sub-surface flow and tile drainage flow) and pesticide properties. However they concluded that point source of pesticides could be reduced by improving sprayer filling and washing areas to reduce the amount entering a buffer zone. Another idea has been to examine whether an artificial wetland can be used as a wastewater treatment dissipating pesticides by adsorption on substrates, for example vegetation, straw, sediments and clay (Tournebize et al., 2011). An understanding of the fate of pesticides is essential for rational decision-taking regarding their authorisation.

Studies by de Snoo and de Wit (1998) confirmed that the amount of pesticide deposited in ditches (Fig. 6.2) was affected by the choice of nozzle and wind speed. They concluded that with a 6 m buffer zone no deposition was recorded in a ditch when the wind speed was 4.5 m/s, and therefore having unsprayed crop edges offered a good way of protecting aquatic ecosystems. For some pesticides, wider buffer zones are required. Byron and Hamey (2008) concluded that while different methods of risk mitigation can be used, a system, such as LERAP provides flexibility, is easily implemented by the user and is enforceable.

In the UK, farmers were not keen to lose 6 m around the edges of their fields. In view of developments in spray technology, it was decided that a narrower buffer zone would be acceptable if the method of application and/or dose of pesticide applied was adjusted. This led to the Local Environmental Risk Assessment for Pesticides (LERAP) being developed (Gilbert, 2000) (Table 6.1). Farms have as many as 75% of their fields alongside water courses, so adoption of LERAP is important. In arable crops, many farmers have adopted the use of LERAP 3* nozzles, usually air

(a)

Fig. 6.2 (a) Measuring drift into a ditch.

(b)

(c)

Fig. 6.2 (*Continued*) (b and c) Measuring drift in Holland (Photograph by Jan van den Zande).

induction nozzles, to reduce the potential drift and allow the width of buffer zones to be reduced. In practice, farmers tend to use the 3* nozzle over the whole field, whereas they could use a more suitable nozzle over much of the crop and change to the coarser spray for the last downwind swaths across the field nearest to a watercourse or ditch. Drift can be reduced by shutting off nozzles when close to the field boundary (Fig. 6.3), ensuring the boom is not set too high (Fig. 6.4) and checking the wind speed to ensure it is within the recommended limits (Fig. 6.5), or changing

Table 6.1 LERAP calculations to determine minimum width of buffer zone (m) from top of the bank for arable crops, with or without 3* nozzle rating

Spray equipment	Standard equipment				Using LERAP 3* nozzles			
Dose rate	Full	3/4 rate	1/2 rate	1/4 rate	Full	3/4 rate	1/2 rate	1/4 rate
Size of stream								
<3 m	5 m	4 m	2 m	1 m	1 m	1 m	1 m	1 m
3–6 m	3 m	2 m	1 m	1 m	1 m	1 m	1 m	1 m
>6 m	2 m	1 m	1 m	1 m	1 m	1 m	1 m	1 m
Dry	1 m	1 m	1 m	1 m	1 m	1 m	1 m	1 m

Fig. 6.3 Avoiding spray from nozzles at the end of a boom to reduce drift of spray into hedge (Photographs courtesy of Hardi International).

the nozzle (Fig. 6.6). There is a trend to increase tractor speeds as well as increase boom width, but more air turbulence is created behind the tractor at the faster speeds (Fig. 6.7).

Where lower doses are used, they must be used over the whole treated area. In the UK, air is usually sufficiently moist ($\Delta T<7°C$), but in arid areas low humidity can increase evaporation from droplets. As droplets shrink in volume, they are more likely to remain airborne and drift further (Parkin et al., 2003). Actual drift will also be affected by the formulation and concentration at which it is applied (Butler-Ellis and Bradley, 2002). Herbst (2003) using vertical collectors in a wind tunnel to calculate a drift potential index (DIX), reported that using a particular air induction nozzle, the drift potential

Fig. 6.4 Checking the height of boom to avoid causing too much downwind spray drift (Photographs courtesy of Hardi International).

Fig. 6.5 Checking wind speed as to suitability for spraying (Photographs courtesy of Hardi International).

was 50% compared with a flat fan nozzle (F110/1.2/3.0), but when the herbicide glyphosate was used, the drift potential was only reduced by just over 25%. This may have been due to the formulation affecting the surface tension of the spray liquid and thus the breakup of the liquid sheet.

Fig. 6.6 Comparison of spray drift with conventional flat fan nozzle and reducing drift with different nozzle (Photographs courtesy of Hardi International).

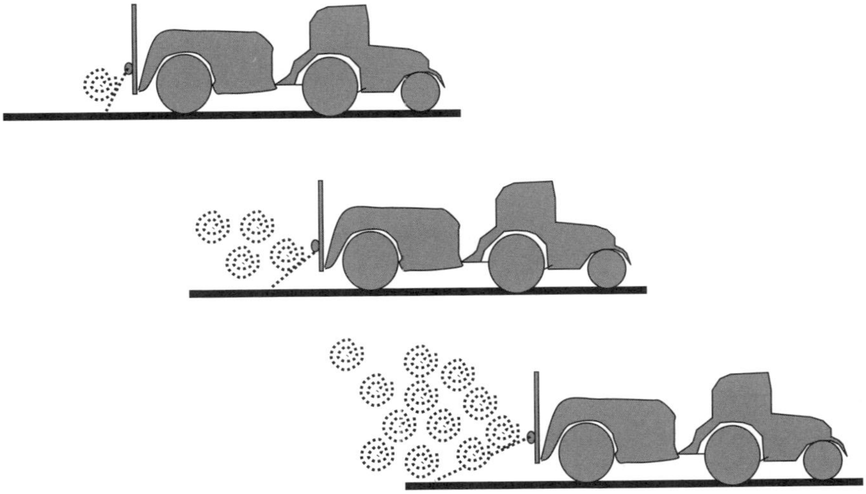

Fig. 6.7 Higher tractor speeds create more turbulence, causing (potentially) more spray rift behind the tractor.

In 2011, the UK introduced an interim measure to have larger aquatic buffer zones of up to 20 m, for horizontal boom sprayers applied on a crop basis rather than a product basis, unlike the current LERAP scheme. The wider distances cannot be reduced under the LERAP scheme. Similarly in Spain in relation to spraying citrus, Cunha et al. (2012) suggested that some pesticides may pose a risk to aquatic organisms, even with a 20 m buffer zone, and also to bees, and adult and child bystanders.

Thus the width of a buffer when using certain pesticides may be set much wider. One example is when chlorpyrifos is applied, even with LERAP 3* nozzles, the no-spray buffer must be 20 m wide to protect water courses, in contrast to 1 m for a dry ditch. Stewardship programmes have been encouraging farmers to accept wide buffer strips in preference to a pesticide being banned. Adoption of buffer zones would be greater if there was a financial incentive, for example by allowing narrow strips to be eligible as set-aside areas.

In Germany, maps have been made of rivers and their tributaries so that using a GIS-based decision support system with a graphical user interface, a farmer can determine where he can treat his fields without infringing the regulations to protect water (Ganzelmeier, 2005; Ropke et al., 2004) (Fig. 6.8).

A similar LERAP system operates for orchards, but as the risk of drift is greater, the minimum buffer zone is 5 m, even with tunnel sprayers. Those treating orchards are in a more difficult situation as spray is directed upwards into and over tree canopies. Complex interactions between the air passing through and over the orchard with the airflow from the sprayer occur so droplet movement is much more difficult to predict compared with field crops. In some cases droplets travel in the opposite direction to the cross-flow (Farooq et al., 2001). Typically an undirected axial fan has been used on orchard sprayers and for some pesticides very wide buffer zones were needed. In bioassay tests, mortality of 10% was recorded at about 50 m downwind of orchard spraying (Davis et al., 1994a). Many orchards are protected by wind-breaks of alder and other plants, and these barriers act as filters to reduce downwind drift. Such a filter needs to be sufficiently porous to allow air flow yet have sufficient foliage to collect the spray droplets. Thus the efficiency in terms of filtering improves in late spring, early summer as the amount of foliage increases. In practice a farmer needs to be careful which pesticides are used to avoid killing beneficial insects that survive in the wind-breaks. If a wind-break is too solid air-flows tend to go over the top of the wind-break or hedge. In one study there was a sudden decrease in deposition in the shelter of the hedge, followed by a gradual increase over the next 15 m, a distance equivalent to ×9 the hedge height (Davis et al., 1994b). While a hedge and vegetation along streams will protect a water course downwind, smaller droplets carried by the air-flow will also be filtered on insects and vegetation over a longer distance.

Some sprayers have been adapted with shields to minimise drift (e.g. Sidahmed et al., 2004), while others have enclosed the trees in a mobile 'tunnel' while spraying. Growing smaller trees in trellises facilitates some of the newer multi-row sprayers. In some situations with flat land, use of a 'tunnel'; sprayer enables spray that penetrates through the crop canopy to be collected on the tunnel wall and re-circulated (Fig. 6.9). While drift can be reduced, uniformity of deposit on tree canopies is more difficult to achieve unless the nozzles are correctly adjusted relative to the tree profile (Planas et al., 2002).

In the United Kingdom, a system using pictograms for pesticide dosage adjustment in relation to the crop environment (PACE) for apple orchard spraying with axial-fan equipment takes account of different tree sizes and canopy density (Fig. 6.10) (Cross et al., 2004). Reductions in the pesticide dose can be taken into account as a factor when using a LERAP and may allow a reduction in buffer zone widths.

In practice spray drift is not the main source of contamination in water. In many cases the most serious cases of pesticide pollution are due to spillages, especially of the undiluted pesticide, run-off from surface

(a) Minimum distance of an agricultural field to the nearest water body
Converting surface water and field geometry from ATKIS in a GIS-raster format with a pixel (cell) resolution of 5 meter

(b) Minimum distance of an agricultural field to the nearest water body
Calculating for each cell the distance (Euclidean distance) to the nearest water body

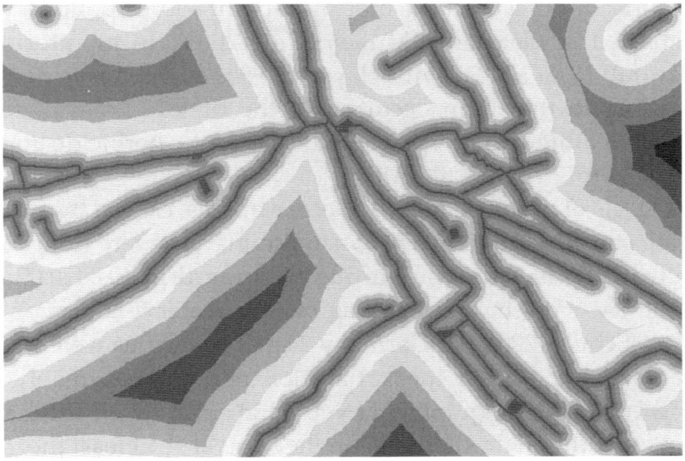

Fig. 6.8 Using GPS and GIS in Germany to protect rivers (Ganzelmeier, Germany). Different sources of pesticide affecting water sources, (b) calculating distances from the water courses and (c) position of tractor in relation to wind direction and strength.

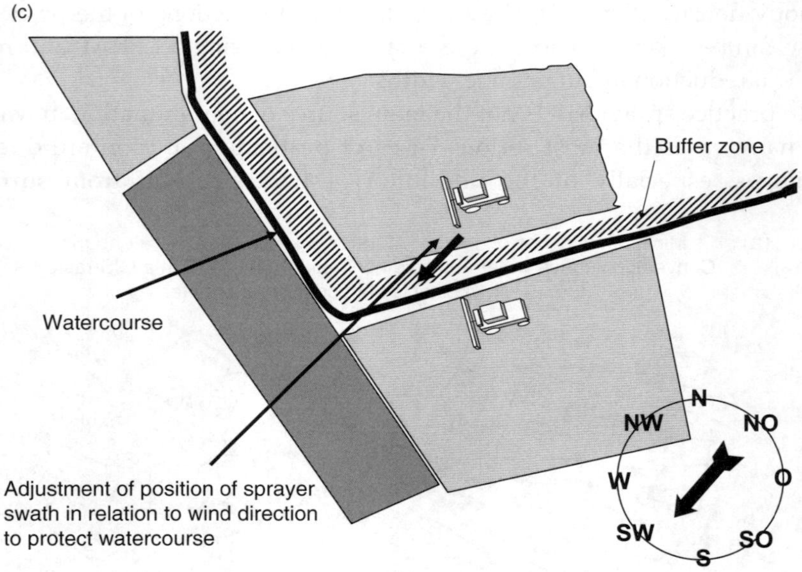

Fig. 6.8 (*Continued*)

deposits and the washing out of equipment, especially if this is done on hard surfaces (Figs. 6.11 and 6.12). One example of these problems is the use of herbicides in urban areas to keep gutters free of weeds along the edges of roads, although washing out sprayers on concrete farm yards has also led to chemical being washed into drains. This is most obvious if heavy rain occurs soon after an application. Single rainfall events, which resulted in run-off caused the most non-point source pollution in a catchment studied in Germany (Muller et al., 2003). In monitoring a single drain outfall from a field with clay soil, 99% of the pesticide, sulfosulfuron, loading to the drain occurred in the first 12.5 mm of flow within 14 days of treatment and represented 0.5% of the herbicide applied to the 7.7-ha site (Brown et al., 2004a).

Peak levels of pesticide in river water samples, caused by surface run-off following storms involving 6.8–18.4 mm rain/day, were from 2 to 41 times higher than the levels recorded due to spray drift (0.04–0.07 μg/l), even when the river discharge rate was much greater 7–22.4 m^3/s instead of 0.28 m^3/s (Schulz, 2001). In Germany 24 g of pesticides were found in each farmyard run-off during the application period, presumably caused by cleaning the spraying equipment (Neumann et al., 2002). This has led to the need for an additional water tank on the sprayer so at least the inside of the tank is washed in the field and the washings used to spray the last part of a field. A new International standard for sprayer washing (ISO22368 parts 1–3) has also been published.

(a)

(b)

Fig. 6.9 Tunnel sprayer enclosing nozzles to reduce drift (a) in vines and (b) in apples (Photograph by Greg Doruchowski, Poland).

Similar rapid loss of herbicides following application to the kerbside of roads during the first rainfall event was reported by Ramwell et al. (2002), and of an insecticide applied to turf (Armbrust and Peeler, 2002), but not from railway tracks treated with herbicide (Ramwell et al., 2004). In fact the

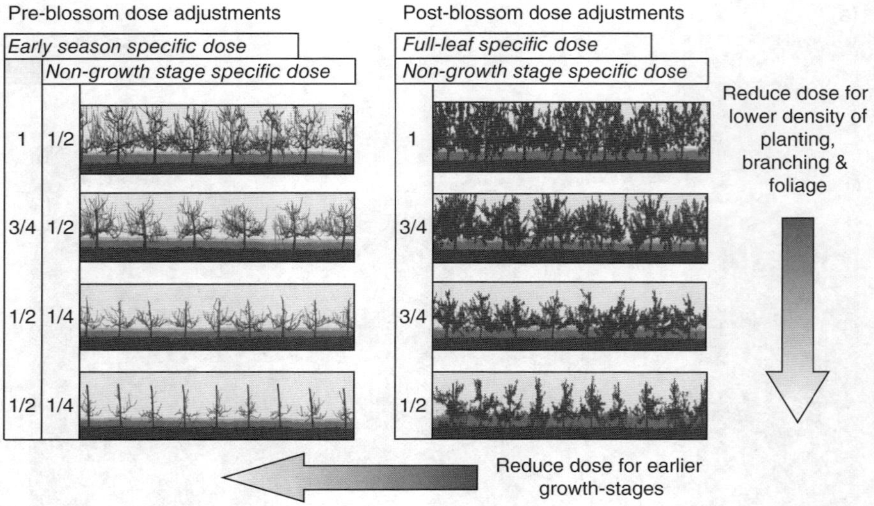

Fig. 6.10 Using PACE to decide on spray dosage (Photograph by Peter Walklate, Silsoe Research Institute).

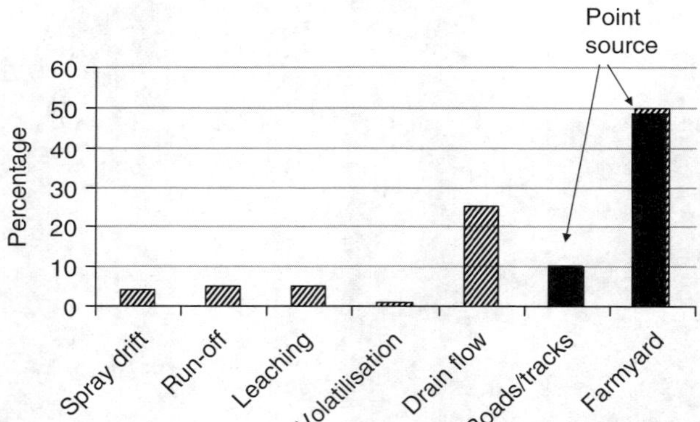

Fig. 6.11 Sources of pesticide pollution in water.

impact of pesticides in urban areas can be higher than neighbouring agricultural areas, thus the highest concentration of diuron in a French catchment area was 8.7 μg/l due to its application on hard surfaces (Blanchoud et al., 2004), a level which is far in excess of the EU drinking water standards. In Germany, non-agricultural use of pesticides contributed more than two-thirds of the pesticide load in tributaries and at least one third in the River Ruhr (Skark et al., 2004). Even quite low concentrations of pesticides, such as greater than 0.01× acute toxicity to *Daphnia* (49 h LC_{50}), affected the macro-invertebrate community structure in agricultural

Fig. 6.12 Pesticide spilt on the ground removed for safe disposal (Photographs courtesy of Hardi International).

streams due to run-off (Berenzen et al., 2005a). This emphasises the need for safety factors in assessing registration of pesticides.

In Denmark, a pesticide action plan was introduced in 1986, and subsequently a tax was imposed on pesticides in 1996. It was thought that if farmers used 30–50% less pesticide, it would not have an impact on GDP. The main index used was the Treatment Frequency Index (TFI), which showed relatively little change as farmers endeavoured to maintain production. More recently it has been suggested area load (AL load per hectare) would be a more satisfactory index as it can be used to pinpoint where most pesticide is being used, thus fungicides on potatoes have a higher AL. Interestingly, in France a study showed that reducing pesticide use by 30% could be possible without reducing farmers' income (Jacquet et al., 2010). The problem for taxing a pesticide is how to choose a system that has the desired effect in the field. Legislation and registration authorities already can restrict which pesticides may be used and improved application can enable farmers to use a dose lower than that on the label very effectively.

In California, it had been suggested that a 2.1% tax on wholesale value of pesticides should be increase to 10% for three years to generate extra funds to support educational programmes for growers on how to reduce the volume of, or eliminate, pesticide run-off. Those growers who enrolled for training would get an incentive by receiving a rebate that would compensate for the increased tax.

When rain washes deposits from the crops, apart from reduced efficacy, it also contributes to water pollution. Schutz (2001) reported rainfall-induced run-off from orchards following a storm that precipitated 28.8 mm of rain.

Increased concentrations of several pesticides were detected, with some extremely high levels that exceeded national water quality standards. The effects persisted for about 3.5 months, thus illustrating that a short-term exposure has the potential of longer term effects. A simple model to predict pesticide run-off in many streams on a landscape level has been proposed where limited data is available (Berenzen et al., 2005b). A concern is that where farmers adopt air induction nozzles to reduce spray drift, there could be more endo-drift on the soil surface to contribute to run-off. Run-off following rain is also reported to be increased where crops are covered by plastic covers or mulches (Arnold et al., 2004). Run-off rates of 8–22% of nine herbicides from rice paddies studied in Japan showed a correlation with the octanol–water partition coefficient log P_{ow} rather than the water solubility of the herbicides (Nakano et al., 2004).

In developing countries, there is little relevant data, although there is increasing public concern increasing with respect to pesticide pollution of the environment and of drinking water resources. In Vietnam, in addition to routine monitoring of certain pesticides in the Mekong delta, a survey aimed at household level pesticide management revealed a wide range of pesticide residues were present in water, soil and sediments throughout the year. This finding was of particular concern as water for drinking is often extracted from canals and rivers by rural households, but preparation of the surface water prior to consumption by flocculation and boiling is insufficient for the removal of pesticides and boiling can actually increase the concentration of non-volatile pollutants (Toan et al., 2013).

Tariq et al. (2003) detected several locally used pesticides in open wells in Pakistan, that were used as rural water supplies. In this particular case the maximum concentration levels established by the US EPA were not exceeded. However in India, Sankararamakrishnan et al. (2004) reported high concentration of organochlorine and organophosphorous pesticides in surface and ground water, with the concentration of malathion being much higher than the EC water quality standards. Ecological monitoring methods for the assessment of the impact of pesticides in the tropics have been published (Grant and Tingle, 2002) together with guides for practical field assessments.

Pollution of water due to pesticides also needs to be put into the context of other pollutants, some of natural origin. Sampling directly under areas of bracken, for example, shows that a carcinogenic water-soluble substance ptaquiloside can be detected at 7 μg/l at a depth of 90 cm. Fortunately, it is unstable under both acidic and alkaline conditions and will transform to pterosin B, which is harmless (Lars et al., 2003).

In the UK, water is sampled in many locations and at frequent intervals to meet the EU Directives on water quality, the data being kept at the Toxic and Persistent Substances (TAPS) Centre. The Environment Agency publishes a summary of the data, which shows where surface and ground waters contain pesticide concentrations that exceed the 0.1 μg/l limit. Pesticides that exceed the limit are usually herbicides, from large-scale use on farms, but exceedances may also be related to urban use. Guidelines are also available, for example for those dipping sheep and for safe disposal of a quantity of water, in this case with pesticide from the dip (Anon, 2001). A number of computer models have been developed to assist with assessing the amount of pesticide in water. SWATCATCH is one model that can simulate maximum concentrations of pesticide at different times in surface waters (Brown et al., 2002).

In the USA water is similarly sampled under the National Water-Quality Assessment (NAWQA) Program of the US Geological Survey. Between 1993 and 1995, data was obtained from 2227 sites (wells and springs) sampled in 20 major hydrologic basins across the United States. Six herbicides were detected in shallow ground water, but overall more than 98% of the detections during the NAWQA investigations were at concentrations of less than 1 μg/l, and this standard was exceeded at fewer than 0.1% of the sites. All of these exceedances involved atrazine. During the survey, two or more herbicides could be detected only at fewer than 20% of the sites sampled.

An inevitable outcome of contamination of water is possible effects on populations of fish and other aquatic organisms, especially the crustacea and molluscs. The activity of the botanical insecticide rotenone was first noted when used traditionally to stun fish, which were then easy to net. Not surprisingly, major fish kills followed use of highly toxic insecticides in lowland irrigated rice paddies. This is a particular problem where fish farms are sited alongside irrigated land as in many parts of Asia. Endosulfan is particularly toxic to fish, such as Nile tilapia (*Oreochromis niloticus*) (Cagauan, 1995) and can persist in paddy water and soil for up to 73 days after spraying, although microbial activity does cause degradation.

In contrast to agricultural use where whole fields were treated, studies on the impact of DDT used to control tsetse flies in Zimbabwe, where deposits were localised to tsetse resting sites, resulted in no indications of fish kills in the Kariba area, although residues of DDT and metabolites were higher in fish sampled in sprayed areas with seasonal rivers flowing into the Zambezi, compared to unsprayed areas. These fish were considered to be a source of contamination for fish-eating birds, such as the fish eagle (Douthwaite and Tingle, 1994). Detailed ecotoxicological studies were also conducted in Botswana where aerial sprays of a low dose of 6–12 g/ha endosulfan was applied as an aerosol of droplets smaller than about 70 μm against tsetse flies in the Okavambo swamp (Fox and Matthiessen, 1982). Careful application of sequential drift sprays resulted in concentrations of 0.2–4.2 μg/l in water 6–9 h after spraying. Apparent mortality of fish varied from 0 to 60%, but high kills were sporadic and possibly due to a leak from the equipment. Overall only some small fish were affected with an average of 0–4% mortality per cycle depending on the species. Later using deltamethrin was safer for fish, although it affected more arthropods in tree canopies. This is due to pyrethroid insecticides being adsorbed onto soil particles suspended in the water. As indicated earlier some studies have been done in Europe to determine whether aquatic ecosystems can be protected by artificial wetlands positioned to receive tile drainage discharge, and be effective pesticide pollution remediation tools (Tournebize et al., 2013).

In contrast to avoiding pesticides getting into water, there are some situations where treatment of water with pesticides is needed. These include the application of insecticides as larvicides to control the immature stages of important vectors of human diseases, notably Anopheline mosquito vectors of malaria and other mosquitoes as well as *Simulium* spp. the black fly vector of onchocerciasis that causes infected people to become blind. In some cases herbicides used to control algae and water weeds, such as water hyacinth will also require direct application to water surfaces. In most cases, preference is now given to *Bacillus thuringiensis israelensis* (Bti), although the organophosphate insecticide temephos, pyrethroid permethrin and some insect growth regulators such as pyriproxyfen are used.

Recently, there has been concern that some pesticides and their metabolites may remain in the environment bound to soil and are not extracted by the usual chemical processes. These bound residues may be so tightly bound that they are essentially not available, but some have wondered that if the load of these chemicals built up, a time may come when they may be released. The risk of future problems is difficult to assess, but for single additions of individual pesticides, their binding to the soil seems to provide an environmental solution to the problem (Barraclough et al., 2005). The situation is not so clear where different compounds are involved with multiple residues.

Protecting vegetation

Reduction in drift is also crucial to protect plants downwind of treated areas. Much of the attention to drift was initiated when horticultural crops were damaged by a volatile herbicide had been applied to control broad-leaved weeds in cereals (Elliott and Wilson, 1983). Koch et al. (2004a) has illustrated by using paraquat herbicide that the effect of the large droplets that sediment rapidly is clear-cut and limited, although there can be a short downwind displacement of the spray. However, the airborne droplets, affected by surface friction and turbulence cause trails of drift that influence their distribution and subsequent effects downwind. With paraquat, further downwind effects are shown by individual droplets scorching the leaves. This illustrates the difference between laboratory studies using low doses of pesticide in relatively high volumes to assess effects on non-target species, with the reality of a patchy distribution of a very low volume of liquid in small droplets, containing a higher concentration of pesticide (Koch et al., 2004b). In consequence an alternative to the laboratory investigations was suggested by Koch and Weisser (2004). The ability of plants outside crops to survive complete kill and still produce seed is crucial, not only from the point of view of plant biodiversity, but also as a source of seeds for birds and other non-target organisms.

Exposure to non-target terrestrial organisms and the plants on which they can forage has been estimated using the standard drift data developed Ganzelmeier et al. (1995), but as this is mainly concerned with deposition on the ground at different distances downwind, it can significantly underestimate the true exposure (Lane and Butler-Ellis, 2003). Most drift studies were always over open ground, but edges of fields have hedges and other vegetation which affects the air flow (Figs. 6.13 and 6.14). Individual plants in a hedge not only provide a vertical barrier, but will filter out some of the droplets that are airborne and still large enough to impact on leaves and stems, rather than continue carried in the air flow. Unfortunately non-target organisms within the hedge can be affected by the drift collected there.

Studying the effect of drift in relation to a hedge and effect of gaps in a hedge, Davis et al. (1993) used various sampling surfaces to detect a fluorescent tracer and also assessed the impact of the herbicide MCPA on seedlings of Ragged-robin (*Lychnis flos-cuculi*). Results confirmed similar bioassays with an insect that the shelter effect immediately behind the hedge, but 13 m behind the hedge there was little or no difference with results obtained behind a gap.

Fig. 6.13 Measuring drift adjacent to a hedgerow (Photograph by Paul Miller, Silsoe Research Institute).

(a)

(b)

Fig. 6.14 (a) A hedgerow can filter downwind drift if it is sufficiently porous to avoid wind taking the spray up and over the hedge. (b) Wild flowers in field border (Photograph by Graham Matthews). (*See insert for color representation of the figure.*)

As a follow up of studies in relation to protecting water in ditches, de Snoo and van der Poll (1999) showed that alongside edges of wheat fields that were not sprayed to leave a buffer zone, the diversity and cover of dicotyledons increased, enhancing the floristic value of the vegetation. Similar changes were not significant alongside potato or sugar-beet, probably because of differences in herbicide use. De Snoo (1999) pointed out that from a farming perspective, it is important to have flexibility in the width of unsprayed crop edges. In Canada a vegetated 10 m field margin provided protection from herbicide drift into a wetland area under wind conditions normally acceptable for spraying, but in high winds this needs to be extended to 20 m unless there is also a windbreak with a porosity of 25% (Brown et al., 2004b). Some porosity is essential otherwise the wind will take any droplets up and over the barrier. Using a probabilistic model to assess risk to organisms in an aquatic environment with chlorpyrifos as an example being sprayed on top fruit, 5% of TER values will be less than 1 even with a buffer zone of 80 m, thus the EC_{50} for randomly selected arthropod species will be exceeded after 5% of spray events with this wide a buffer zone (Crane et al., 2003). Wennecker and Van de Zande (2008) reported that when spraying an orchard with a natural windbreak with broad-leaved trees, the risk of spray drift is high during the early part of the growing season, depending on the extent leaves have developed.

Using LIDAR equipment, Miller and Stoughton (2000) showed that when an aerial spray was applied to the edge of a hardwood forest, small droplets were dispersed in the atmospheric boundary layer. The implication being that even with well-conducted spray operations a small amount of pesticide will be widely dispersed. However where bracken has to be controlled by applying a herbicide, asulam, aerially in inaccessible areas, such as hillsides, on which a farmer wishes to graze sheep, drift from 'raindrop' drift-reducing, hollow-cone nozzles on a helicopter was mainly limited to 35 m downwind. Drift beyond 35 m of the treated area was similar to the drift from ground equipment. On the basis of the trials a 50 m buffer zone was approved by the Environment Agency in the UK (Robinson et al., 2000).

Studies of drift into woodland alongside a sprayed field (Fig. 6.15) indicated that penetration depended on the peripheral vegetation and to some extent wind speed, but where vegetation was low, spray droplets could penetrate 10 m, with the highest concentrations confined to within 5 m of the spray boom (Gove, 2004; Gove et al., 2004). Needle-like foliage in windbreaks can capture 2–4 times more spray than broadleaved foliage (Ucar et al., 2003). Thus where 1% of applied herbicide drifts that far into woodland, the most sensitive plants can be adversely affected. Surveys of ancient woodland margins in Kent showed that species richness and abundance was least in margins alongside arable land compared to unimproved grassland (Gove, 2004).

Fig. 6.15 Sprayer alongside a woodland. Photograph by Benedict Gove.

Like crops, non-target vegetation will also vary in its susceptibility to herbicides. Most broad-leaved plants are to some extent susceptible to herbicides, which are used in cereals, such as wheat, to control broad-leaved weeds. Similarly herbicides designed to control grass weeds will also adversely affect natural grass vegetation. With this variability, OECD has suggested that the activity of new herbicides should be evaluated on 6–10 species, including both mono- and dicotyledon species, in addition to crop species when assessing effects of pesticides on plants in terms of preserving biodiversity. Protecting rotational fallows is important as in these areas species richness of flowering plants increases can be mirrored by increases in the abundance of insect pollinators (Kuussaari et al., 2011). Loss or reduction of key plants can result in fewer overwintering sites and can lower populations of natural enemies in early spring, for example tussocky grasses are important for aphid parasitoids (Thomas and Marshall, 1999).

Apart from effects on vegetation around treated fields, farmers must also be concerned with the possible effects of a persistent herbicide on any following crops, or whether a pesticide application will affect an adjacent crop, if inter-row or strip cropping is practised. By using a coarser spray to reduce airborne drift, there is always a possibility of increasing endo-drift, so the choice of pesticide and method of application do need to be considered carefully, where there is a greater diversity of crops.

Protecting pollinators

Of major concern has been the decline in bee populations in countries, such as the USA, where Colony collapse disorder has been a key driver. Declines are due to many factors, including the presence of a mite *Varroa destructor* in bee hives and viruses transmitted by the mites, but the main blame has been directed at insecticides, particularly the neonicotinoids. These systemic insecticides have been used extensively as seed treatments and detected albeit at extremely low levels in pollen and nectar. In Germany a high dosage of clothianidin was applied to maize seed to control corn rootworm *Diabrotica vergifera*, but dust produced during the planting of the treated seed drifted onto adjacent flowering plants and resulted in exposure of foraging honeybees (Marzaro et al., 2011). This resulted in over 11,500 colonies of honeybees being poisoned in Spring 2008 (Pistorius et al., 2009) This incident in Germany led to pressures by the media and NGOs and resulted in the EU Commission imposing a 2-year moratorium on the use of clothianidin, imadacloprid and thiamethoxam on the seeds of bee-attractive crops or crops drilled when flowering crops may be adjacent (EU Commission, 2010).

Subsequently the formulation of the insecticide was improved and modifications to the application method reduced the emission of dust (Foque et al., 2014a and 2014b; Nuyttens et al., 2013). In the UK where a lower dose of insecticide was used on treated seed and with different sowing equipment, bee mortality due to neonicotinoids was not reported. In France spraying during daylight when bees are foraging is also not permitted. In Australia where bees were not affected by the Varroa mite, the same insecticides have continued to be used.

The seed treatment was considered to be the most acceptable environmentally and the least likely to affect bees as the amount that remained in the plant by the time a crop was flowering was extremely low. The greater danger was if other insecticides were applied as sprays to a flowering crop. In Canada a large-scale field experiment to determine whether exposure to clothianidin seed-treated canola (oil seed rape) had any adverse impacts on honey bees suggested that exposure to canola grown from seed treated with clothianidin posed a low risk to honey bees (Cutler et al., 2014).

Interestingly the survival of bees in Europe during the 2013–2014 winter was much better, although neonicotinoid seed treatments were still used in the autumn of 2013. Initial data showed that the overall proportion of colonies in 21 countries lost was 9%, the lowest since the international working group started collecting data in 2007 (COLOSS Press release July 2014). In the UK, the government now has a National Pollinator Strategy, outlining a 10 year plan to help pollinating insects to survive and thrive, especially by restoring habitats favoured by the insects (Anon, 2014a, 2014b).

Protecting birds

The Royal Society for the Protection of Birds has for many years been concerned at the decrease in a number of bird populations in the UK. The use of pesticides has been blamed although it has also been recognised that many other changes in British agriculture have taken place over the last 50 years. The removal of hedgerows to allow more efficient use of large equipment is one of the most significant changes as it removed habitats suitable for many species and reduced food supply for some insectivorous birds. However it is clear that pesticides can affect bird populations in several ways. One is a direct effect of poisoning, secondly ingestion of insects or vegetation that have been sprayed can have an effect, while the removal of vegetation can decrease populations of herbivorous insects that provide the food of certain bird species.

Overall effects of different pesticide types are shown in Table 6.2. Direct effects due to sprays are not expected unless a highly toxic insecticide is sprayed directly on the birds. In Africa, certain weaver birds, *Quelea*, are known to destroy cereal crops ready for harvesting, so large colonies of these birds have been sprayed with the OP insecticide fenthion, when they congregated to roost at night. While this may save crops in fields in the immediate vicinity of the roost, it has never had any impact on overall populations of this bird.

In the 1950s, treating cereal seeds with an organochlorine insecticide, for example dieldrin, to protect them from soil pests, including wheat bulb fly, led to the first bird mortalities, when hungry birds ate spring-sown seeds on or close to the soil surface. Rapid realisation of the cause of bird deaths led to the banning of sowing treated seed in the spring, when few other sources of seeds were readily available after the winter. Later the consequence of predatory birds, such as the peregrine falcon, accumulating

Table 6.2 Effects of different classes of pesticides on birds (Adapted from Newton, 1998. © Elsevier)

Pesticide type	Acute toxicity	Persistence	Bio-accumulation to birds
Insecticides	Very low	Low	Very low
Natural pyrethrins			
Organochlorine (e.g. DDT)	Low	Very high	Very high
Cyclodienes	High	High	Very high
Organophosphates	Very high	Low-moderate	Low–moderate
Carbamate	Very high	Low-moderate	Low–moderate
Pyrethroids	Low–moderate low	Low	
Neonicotinoids	Low	Low	Low
Insect growth regulators	Very low	Moderate	Low
Fungicides	Low	Low	Low
Azole fungicides			
Herbicides	Low	Low	Very low
Other (e.g. paraquat)	Moderate	Low	Very low

these persistent insecticides, became more evident with a rapid decline in their populations from 1955. Noting an abnormally high incidence of egg breakages in the falcon nest (eyries) led to the study of egg shell thickness (Radcliffe, 1967, 1970; Cooke, 1973). Overall insecticide residues in many species of birds that resulted in a reduction of egg thickness of about 17% or more was the cause of egg breakages occurring under natural conditions (Fig. 6.16) (Newton, 1998). In North America, Hickey and Anderson (1968) were the first to report similar eggshell effects. Despite clear evidence of the adverse effect of organochlorine insecticides, 260 000 acres of wheat were aerially sprayed with endrin as late as1981 in the USA contaminating many bird species with residues (Metcalf, 1984). Persistent organochlorines can still be detected in bird populations, for example in vultures in South Africa (van Wyk et al., 2001) and in passerines in North America (Bartuszevige et al., 2002). In Australia, a technique of flushing stomach contents of birds near or away from areas of insecticide use (cotton farms) was used to detect positive residues, including DDT and endosulfan in 90% of the birds sampled (Sanchez-Bayo et al., 1999).

The decline in grey partridges in the UK (Potts, 1986) was due to herbicide sprays removing the weeds on which the survival of a number of insect species depended (Table 6.3). It was the lack of these insects, which deprived the young partridge chicks of food in the early stages of their development. Thus poor survival of the chicks resulted in the population crash. Farmers

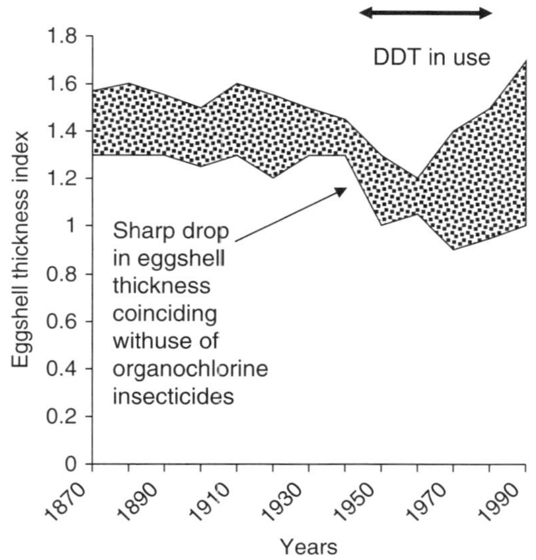

Fig. 6.16 Indication of the shell thickness index of British sparrowhawks, *Acipiter nisus*, from data collected at Monks Wood Research Station. Upper and lower indices are shown. Shell index measured as shell weight (mg)/shell length × shell breadth (mm). Shells became thinner abruptly after 1947 with the first widespread use of DDT and dieldrin; this was followed by recovery following restrictions on the use of these pesticides. Adapted from Newton 1998.

now can leave a strip around their fields, which is not treated with herbicides. This has developed into the concept of conservation headlands to foster partridges, pheasants and other game birds. Within the field, growing a herbicide tolerant row crop can enable weeds to be controlled intra-row and delay weed control in the inter-row to allow herbivorous species of insects access to their food source.

Studies on the yellowhammer (*Emberiza citrinella*) have examined the impact of insecticide sprays on chick survival (Morris et al., 2005). As these birds also depend on an abundance of invertebrate food for chicks to develop, any insecticide spray at a critical stage in the breeding season will have an adverse effect on their foraging and decrease populations. Numbers of invertebrates are generally higher in the hedge and field margin (Thomas and Marshall, 1999) (Table 6.4).

Several methods can be adopted to encourage bird populations on farms. In addition to conservation headlands, the use of no-spray buffer zones can be developed as 'set-aside' areas. However, these areas need to be managed, as uncared land will revert to woodland. Thus to provide a better habitat for birds, they need to be sown with seeds of grasses and wildflowers to provide pollen, nectar and seeds suitable for the particular bird species, which are to be encouraged in a given habitat. Wild flower mixtures will also encourage butterflies, bumble bees and other beneficial insects (Fig. 6.14b). Several agri-environmental projects are evaluating different approaches to habitat management. In Scotland, it was shown that

Table 6.3 Comparison between plots with or without herbicide on weed, insect and game birds (Newton, 1998. Reproduced with permission of Elsevier)

	Herbicide treated	No herbicide
Weeds		
Number/0.25 m^2	2.1 ± 0.5	6.8 ± 1.0
Total weed biomass	0.9 ± 0.2	0.7 ± 2.3
Percentage weed cover	2.9 ± 0.6	14.2 ± 2.3
Insects		
As bird chick food-items/0.5 m^2	18.9 ± 9.2	67.9 ± 23.4
Brood size		
Grey partridge	7.5 ± 0.8	10.0 ± 0.6
Pheasant	3.2 ± 0.5	6.9 ± 0.5

Table 6.4 Mean number of invertebrates in different positions of fields as indicated by suction samples (Thomas and Marshall, 1999. Reproduced with permission of Elsevier)

Position	Field A	Field B
Hedge	233.0	232.0
Sown field margin	155.8	147.8
Crop edge	84.5	58.3
Field	96.8	76.9

up to 80 times as many birds were recorded where 'game' crops, such as kale and black mustard were grown on set-aside, when compared with nearby conventional crops (Parish and Sotherton, 2004). In Japan, strips >300 m wide favoured bird diversity, whereas strips <50 m wide were unsuitable for feeding by egrets in irrigated rice areas, with some birds hardly occurring at field edges, indicating a need to consider both width and location of strips (Maeda, 2005).

Protection of these areas does require careful thought about when and how pesticides are applied in the adjacent crop fields. The use of certain nozzles, for example air induction nozzles on the last downwind swath adjacent to a 'wildlife strip' can significantly reduce drift into the area, but avoiding a pesticide spray at critical times of the year is also important, especially in relation to the breeding season of birds (Table 6.5). Thus application of certain insecticides and herbicides is better in the autumn on winter sown cereals, rather than applying sprays in the spring and summer. Nevertheless when farmers need to spray at these times, the impact on non-target species can be mitigated, if the least hazardous and more selective pesticide products are used to control specific insect pests or weeds. Application of a reduced rate can also help as some natural enemies detect and avoid pesticide residues. Where the location of specific weeds can be mapped, then patch spraying would also reduce the likelihood of adverse effects in the 'wildlife strip'.

Collins et al. (2002) had shown that strips within crop fields, usually referred to as 'beetle banks', sown with tussock-forming grasses, such as red fescue and timothy grass, conserved ground dwelling carabid beetles, that are important aphid predators. Other studies confirmed a relatively high abundance and number of species of beetles within field margins contribute significantly to invertebrate biodiversity in agricultural landscapes (Woodcock et al., 2005). There are now agri-environment schemes in place across Europe (Tscharntke et al., 2005; Geiger et al., 2010). In England, the establishment of grass margins, established as buffer zones around arable fields, are likely to have the greatest impact on biological control within the now mandatory IPM policy, especially if wild flowers are also present.

Table 6.5 Guidelines of spraying in association with conservation headlands in the UK

Time of application	Autumn spraying	Spring spraying
Insecticides	Only by avoiding drift	Only prior to mid-March
Fungicides	Yes	Yes
Plant growth regulators	Yes	Yes
Herbicides (grass weeds)	Only selective graminicides	Only selective
Graminicides (broad-leaved weeds)	No*	No*

*Some herbicides may be used if approved for a specific weed problem, for example *Galium aparine*.

By 2009, there were more than 58 000 agri-environment scheme agreements, covering over 6 million ha – almost 66% of the agricultural land (Anon, 2009). An examination of cereal aphid control in winter wheat on 14 farms in the south of the UK with contrasting proportions of grass margins showed that levels of aphid control were positively related to the proportion of linear grass margins within 250, 500 and 750 m radii of study arenas (Holland et al., 2012).

Similar studies in the USA recognise the importance of landscape composition by planting floral strips as an important driver of beneficial insect populations and resulting ecosystem services. Coccinellid abundance in soybean was positively related to amount of semi-natural vegetation in the landscape, enhancing aphid suppression (Woltza et al., 2012).

Apart from benefiting beneficial insects, monitoring set-aside fields in Finland from 2001 to 2006 showed that these fields supported 25–40% more bird species and held 60–105% more pairs of birds typical of open farmland in comparison with cereal fields within a similar landscape setting, thus enabling considerable changes in bird populations on the national scale (Herzon et al., 2011).

Overall environmental impact assessments

Most registration authorities examine data on the impact of pesticides on specific key indicator species. The OECD has a project on pesticide terrestrial risk indictors and has published reports at http://www.oecd.org/env/ehs/pesticides-biocides/pesticideriskreduction.htm (pesticide risk indicators). In the UK, the 'Pesticide Forum' publishes an annual report in which details of the various indictors of the impact of pesticides in the environment are described. There are various techniques which can be used to estimate the overall impact of pesticide use in the environment. In 2004, 17 indicators were used taking into account developments with the Voluntary Initiative. Using these indicators, the forum is hoping to improve ways to measure the relationship between use, need and application rates to reduce environmental impact. Some pollution models rely on detailed knowledge of physical, chemical and microbial processes that affect the persistence and movement of pesticides in soil, air and water and may not consider the effect of different organisms within an ecosystem. Some risk assessments consider the fate and exposure to pesticides and attempt to rank effects, while others consider different impacts over the lifetime of a product.

Margni et al. (2002) have advocated the need for a new quantified evaluation of the overall impact of pesticides on the health of humans and ecosystems. The proposed method considers different exposure effects (inhalation, intake via food and drinking water, etc.), transfers, such as soil to water and between water and air, as well as the fate of the pesticides and exposure to them. In developing their approach they have assumed, for example, that following a field application of a water-based pesticide using a boom sprayer, 10% of the spray remains in the air (or returns to the air by volatilisation from foliar deposits). Furthermore, they assumed 85% enters the soil and only 5% is retained on the crop. There are then dilution factors within the soil profile and transfer to surface waters. In subsequent analysis of residues in food (discussed more fully in Chapter 7) they assume that following peeling, washing and processing of a food, the residue remaining is 5% of the tolerance value, that is the MRL. While these assumptions are not always appropriate, they develop characterisation factors that allow an estimate of the impact per kilogram of active ingredient. This then needs to be adjusted depending on the amounts actually applied. Initial evaluation of the technique indicated that impacts on human health, the aquatic ecosystem and terrestrial ecosystem differed between the five fungicides examined. Also in relation to human health, food intake resulted in the highest toxic exposure by $\times 10^3$ to 10^5 compared to drinking water or inhalation.

In the USA a framework for ecological risk assessment was published by the EPA (1992), and this was subsequently updated under the EPA Superfund program which issues updates at http://www.epa.gov/oswer/riskassessment/ecoup/index.htm.

In the UK, a computer-based decision support tool pEMA (Fig. 6.17) (EMA – Environmental Management for Agriculture), estimates risks to a wide range of taxonomic groups in different environmental situations. Methods consistent with the UK regulatory assessments are used but adjusted to take into account the formulation used and local conditions. Pathways along which the pesticide is dispersed in the environment are modelled to estimate concentrations in soil in the field and at its margin, in surface water and groundwater (Brown et al., 2003). Predicted environmental concentrations (PECs) are then combined with toxicological data as toxicity: exposure ratios to facilitate risk assessments to be made. Combining risk indices for individual applications of each active ingredient to form an aggregate score for farm provided an index of the environmental performance or 'eco-rating' for the average field (Hart et al., 2003). pEMA has now become the Pesticide Properties Database (PPDB) – a comprehensive relational database of pesticide physicochemical and ecotoxicological data.

Lewis et al. (2003) provide an overview of the system (Table 6.6), and how it compares with other indicators during the European Project CAPER (Concerted action on pesticide risk indicators). In this collaborative programme, the environmental risk of 15 individual pesticide applications was compared using eight indicators (Table 6.7) (Reus et al., 2002). It assumed all pesticides, except glyphosate were applied as foliar sprays to apple trees using an airblast sprayer. Pesticides included in the project were not necessarily approved in all the countries participating in the project. Surface water, groundwater and soil indicators gave similar rankings, but the

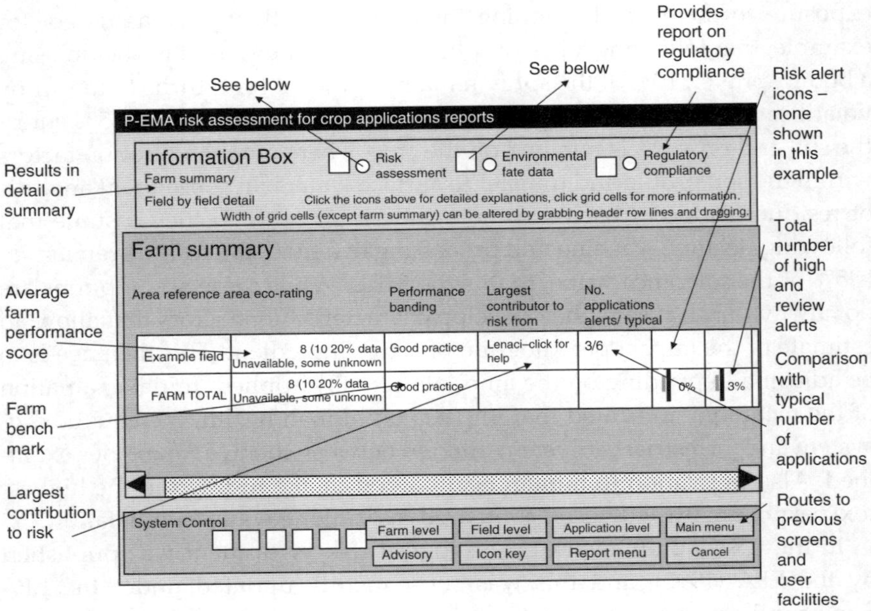

Field by field – risk data

Example field				
Pesticide active applied	Eco-rating mammals	Eco-rating birds	Eco-rating earthworms	Eco-rating algae
Thiram	–45	–44	0	0
Lenacil	0	–15	0	–51
Carbendazim	0	0	–27	0
Flusilazole	–35	–18	0	0
Mean eco-rating	–20	–19	–6	–12

Field by field – fate data				
Example field pesticide active substance	Effective application (a.s.) rate (g.ha)	Annual groundwater conc (ug/l)	Accum. conc. in field: soil (mg/kg)	Accum. conc in margin soil (mg.kg)
Thiram	8.30	0.000 (60%)	0.0111 (100%)	0.0000 (100%)
Lanacil	176.00	0.019 (70%)	0.2347 (100%)	0.0065 (100%)
Carbendazim	16.43	0.000 (100%)	0.0219 (100%)	0.0030 (100%)
Fusilazole	32.85	0.000 (80%)	0.0438 (100%)	0.0060 (100%)

Fig. 6.17 Communicating risk information to the user. Courtesy Kathy Lewis. Selection of specific information Box' on the computer screen provides additional data as illustrated above for risk data and fate data.

overall score for the environment differed. Not unexpectedly a ranking based on 'kilograms of active ingredient' was not correlated with rankings by risk indicators.

In Australia, a Pesticide Impact Rating Index (PIRI) has been used to provide an assessment of the off-site impact of one or more pesticides or

Table 6.6 EMA eco-scores and relative rankings of pesticides used to control aphids and thus virus yellows in sugar beet (Lewis et al., 2003)

	Eco-score with relative ranking in bold						
Pesticide	Mammals	Birds	Earthworms	Average aquatics	Honey bees	Ground water	Overall
Aldicarb	−71 **1**	−100 **1**	no data	−0 **4**	0 **4**	−68 **1**	−45 **1**
Deltamethrin	−28 **3**	0 **4**	0 **2**	−58 **2**	−33 **2**	0 **2**	−20 **4**
Pirimicarb	−58 **2**	−83 **2**	−25 **1**	−35 **3**	−17 **3**	0 **2**	−36 **2**
Imidacloprid	−7 **5**	−35 **3**	0 **2**	0 **4**	0 **4**	0 **2**	−13 **5**
Lambacyhalothrin	−27 **4**	0 **4**	0 **2**	−60 **1**	−37 **1**	0 **2**	−21 **3**

Aldicarb applied as granules, and imidacloprid as a seed treatment, others as foliar sprays. Rating **1** is worst case **5** is best case.

Table 6.7 Pesticide risk indicators evaluated in the CAPER project (Reus et al., 2002. Reproduced with permission of Elsevier)

Number	Risk indicator	Acronym	Country
1	Environmental yardstick	EYP	The Netherlands
2	HD	HD	Denmark
3	SYNOPS-2	SYNOPS-2	Germany
4	Environmental performance indicator of pesticides	p-EMA	UK
5	Pesticide environmental impact indicator	Ipest	France
6	Environmental potential risk	EPRIP	Italy
7	System for protecting the environmental impact of pesticides	SyPEP	Belgium
8	Pesticide environmental risk indicator	PERI	Sweden

land uses on a water resource. The model is based on three components, including the value of the asset that is threatened by the pesticide, the sources of threat, to the asset, and the transport method by which the pesticide can move to the asset or water resource (Kookana and Aylmore, 1994; Kookana et al., 2005). The free software has also been extensively used outside Australia.

One of the problems in risk assessment for organisms in the environment is that they experience fluctuating, pulsed or intermittent exposure to pollutants, in contrast to the simplified design of standard ecotoxicity tests. To have a more robust risk assessment Ashauer and Brown (2013) describe a dynamic simulation using toxicokinetic-toxicodynamic (TK-TD) models that link exposure with effects in an organism and provides a basis for extrapolation to a range of exposure scenarios.

A screening tool to assess potential aquatic toxicity of complex pesticide mixtures by combining measures of pesticide exposure and acute toxicity in an additive toxic-unit model, the Pesticide Toxicity Index (PTI) was developed as pesticide mixtures are common in streams with agricultural or urban influence in the watershed. The PTI is determined separately for fish, cladocerans and benthic invertebrates (Nowell et al., 2014).

Acomposite scoring index (EcoRR ecological relative index) in Australia (Sanchez-Bayo et al., 2002), was evaluated in the context of 37 pesticides that can be used in a cotton development. It was considered that the EcoRR scores reflected the potential risk to ecosystems, as it took account of biodiversity, although it was less dependent on toxicity to sensitive species.

Another approach proposed by Padovani et al. (2004) is an environmental potential risk indicator for pesticides (EPRIP), which is based on the ratio of PEC estimated at local level with short-term toxicity data. It reflects a worst-case scenario but can identify those crops on which pesticide use presents the highest risk to non-target organisms. It does this by taking account of multiple applications, synergistic effects and different formulation types. Thus users can assess different management options.

When these environmental risk assessments are carried out on farms, many other factors also have to be taken into consideration. In assessing the five insecticides shown in Table 6.6, aldicarb granules and imidacloprid seed treatment were prophylactic treatments in areas with a high risk of virus yellows infection. Aldicarb, now withdrawn was preferred, if there is also a nematode problem. However the sugar industry in the UK used an early warning system for aphids (Dewar, 1994), so a spray was only required if and when an aphid infestation occurred. Pirimicarb was a selective foliar spray ideal for aphids, but the farmer sometimes preferred a pyrethroid insecticide for greater persistence and a broader spectrum of activity. Later the imidacloprid seed treatment was favoured as the dose applied to seed was low and no sprays were required. Subsequently the use of this neonicotinoid insecticide was withdrawn by the EU on the assumption the insecticide taken up in the plant was adversely affecting bees. Clearly there is a problem in determining the PEC for seed treatments and further studies are needed to reassess and validate the approach by which it is calculated to reach a consensus on environmental safety (Walters, 2013).

Thus in each crop and agroecosystem, local knowledge is important as well as the overall assessment of ecological impact. Various tools have been developed to assist decision making on farms, for example whether a herbicide can be used (Fig. 6.18).

In South Africa, concern about the occurrence of pesticides in water resources and poor management of water treatment facilities led to the development of a project to assess the risks of current agricultural pesticide use to human and animal health. Four indices were used to prioritise active ingredients applied in excess of 1000 kg per annum, ignoring pesticides used in very low quantities. The indices were (i) the quantity of pesticide used (QI); (ii) the toxicity potential index (TP); (iii) the hazard potential index (HP) and (iv) the weighted hazard potential (WHP) derived from the HP multiplied by the ratio of its use to the total use of all pesticides in the country. The TP which ranked pesticides according to scores derived for their potential to cause five health effects (endocrine disruption,

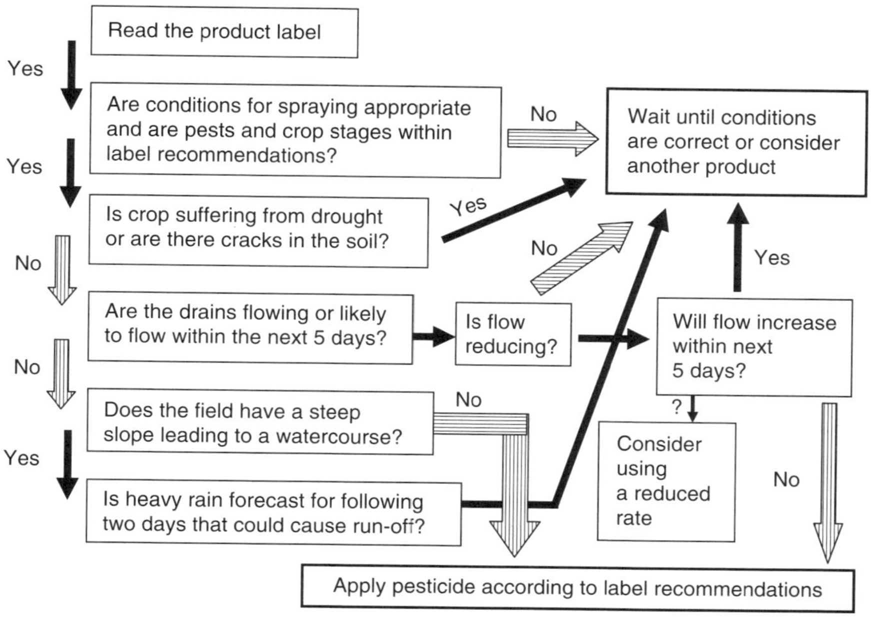

Fig. 6.18 Example of a flow diagram regarding decision on whether to spray a particular herbicide, depending on soil and rainfall. Adapted from Voluntary Initiative.

carcinogenicity, teratogenicity, mutagenicity and neurotoxicity (Dabrowski et al., 2014). It was considered that as crop-specific application pesticide use data was available, the methodology provided important information to develop monitoring programmes, and identify priority areas for management interventions and to investigate optimal mitigation strategies.

In the UK Cross and Edwards-Jones (2006) examined three key indicators of hazard, (i) the Environmental Impact Quotient (EIQ) which rates a pesticide's hazard profile; (ii) the Environmental Impact (EI) which is the product of the EIQ rating and data on actual usage of a pesticide in the UK in a given year; and (iii) EI per hectare which standardised the hazard by dividing total EI in a year by the area of crop grown in that year. They suggested that between 1992 and 2002, the overall hazard posed by the pesticides applied crops declined substantially, as there was a 10% decrease in pesticide usage, an 8% increase in yield per hectare, a 14% decrease in overall EIQ rating, a 15% decrease in EI rating and a 7% decrease in EI per hectare from these trends. These changes were possibly due to use of different pesticides and severity of pests in different years.

Locust control until 1985 had relied on use of dieldrin, which was stockpiled in various African countries. Following the banning of this insecticide, there was a major problem of disposal of the large stocks of obsolete pesticides. FAO set up a programme to cope with this but in anticipation of further locust plagues, a small group was established to advise on suitable insecticides for locust control. The advisory group listed a number of active ingredients, that had been shown to be effective in field trials, and then also provided assessments of environment impact on selected non-target organisms (Tables 6.8 and 6.9) (Anon, 2004). This allowed users a choice between the quick acting organophosphates and pyrethroids for rapid control of locusts in swarms, while able to use either more selective chitin inhibitors against nymphs or a broad spectrum insecticide used at low dose as a barrier treatment. In areas of extreme ecological sensitivity, a biopesticide based on *Metarhizium* was recommended and has been used in Australia as well as in Africa. In the ecologically significant Ikuu wetlands of the Katavi National Park, Tanzania it was applied on 22 000 ha in 2009 and achieved 70% reduction of the population which with other treatments protected 59 800 ha of food crops according to FAO. The biopesticide has also been used in Madagascar during 2014.

However, the greatest task in locust control is to be able to forecast when an upsurge could develop and deploy sufficient resources to control hoppers and incipient swarms rapidly to prevent swarms developing and migrating. The 2003 upsurge was predicted and in October 2003, FAO issued an alert, but funds and resources in West Africa were not forthcoming in sufficient time to stop the upsurge occurring and initially causing crop damage in Mauritania. In contrast in the Sudan and Saudi Arabia similar indications of an upsurge were controlled rapidly. The setting of the Emergency Programme EMPRES to operate in the countries around the Red Sea was one factor that enabled the rapid response. Clearly the more rapidly a potential upsurge can be controlled the less insecticide will be needed. By acting earlier against hoppers, the insect growth regulators and barrier treatments are also likely to reduce environmental impact. The slower acting biopesticide would also be appropriate in situations where crops are not yet at risk.

Pesticide misuse affecting wildlife

Unfortunately the availability of highly toxic insecticides for crop protection has resulted in their misuse to poison and deliberately kill wildlife. Apart from intentional poisoning, populations of non-target animals also can decline, for example the decline in vultures is primarily due to feeding on poisoned carcasses (Ogada, 2014).

In Africa 38 countries have legislation specifically mentioning that it is illegal to use poison for hunting wildlife, but enforcing the law is proving extremely difficult when poachers are seeking elephant tusks and rhino horns while others are obtaining bushmeat for sale. Insecticide treated bed nets distributed to control mosquitoes and reduce transmission of malaria have been used as fishing nets.

Table 6.8 Risk to non-target organisms at verified dose rates against the desert locust (Table 1). Risk is classified as low (L), medium (M) or high (H) (see Table 6.9 for the classification criteria)

	Environmental risk							
	Aquatic organisms		Terrestrial vertebrates			Terrestrial non-target arthropods		
Insecticide	Fish	Arthropods	Mammals	Birds	Reptiles	Bees	Antagonists	Soil insects
Bendiocarb	***M*** [2]	***L*** [3]	***M*** [1]	***L*** [3]	—	***H*** [1]	H [3]	M [3]
Chlorpyrifos	M [3]	H [2]	L [3]	M [3]	***M*** [3]	H [1]	H [3]	—
Deltamethrin	L [3]	H [3]	L [3]	L [3]	L [3]	M [1]	M [3]	M [3]
Diflubenzuron (blanket)	L [3]	H [3]	L [1]	L [1]	—	L [1 Φ]	M [2]	M [3]
Diflubenzuron (barrier) *	L	(H)	L	L	—	L [Φ]	L [3]	(M)
Fenitrothion	L [3]	***M*** [3]	L [3]	M [3]	***M*** [3]	H [1]	H [3]	H [3]
Fipronil (barrier)*	L	M [3]	M [3]	L [3]	M [3]	(H)	H [3]	H [3]
Lambda-cyhalothrin	L [2]	H [2]	I [1]	L [1]	—	M [1]	M [3]	H [3]
Malathion	L [2]	M [2]	L [3]	L [3]	—	H [3]	H [3]	H [3]
Metarhizium anisopliae (IMI 330189)	L [2]	L [2]	L [1]	L [1]	L [2]	L [3]	L [3]	L [3]
Teflubenzuron (blanket)	L [1]	H [2]	L [1]	L [1]	—	L [1 ‡]	M [1]	—
Triflumuron (blanket)	L [1]	H [2]	L [1]	L [1]	L [3]	L [1 ‡]	L [3]	L [3]
Triflumuron (barrier)*	L	(H)	L [3]	L [3]	L [3]	L [1 ‡]	L [3]	L [3]

The index next to the classification describes the level of availability of data:

[1] Classification based on laboratory and registration data with species which do not occur in locust areas.

[2] Classification based on laboratory data or small scale field trials with indigenous species from locust areas.

[3] Classification based on medium- to large-scale field trials and operational data from locust areas (mainly desert locust, but also migratory and brown locust).

* If no field data are available, the risk of barrier treatments is extrapolated from blanket treatments. However, it is expected to be considerably lower if at least 50% of the area remains uncontaminated for a period long enough to allow recovery of affected fauna, and if barriers are not sprayed over surface water. Risk classes are therefore shown in brackets unless the blanket treatment was already considered to pose low risk, and no reference is made to the level of data availability. More field data are needed to confirm that products posing a medium or high risk as blanket sprays can be downgraded to 'L' when applied as barrier sprays.

[Φ] At normal use, diflubenzuron is not harmful to the brood of honey bee.

[‡] Benzoylureas are safe to adult worker bees but some may cause damage to the brood of exposed colonies.

—, insufficient data.

Table 6.9 Criteria applied for the environmental risk classification used in Table 6.2. See text for further explanations

A. Laboratory toxicity data

Group	Parameter	Risk class			References
		Low (L)	Medium (M)	High (H)	
Fish	Risk ratio (PEC/LC_{50})	<1	1–10	>10	FAO/Locustox
Aquatic arthropods	Risk ratio (PEC/LC_{50})	<1	1–10	>10	FAO/Locustox
Reptiles, birds, mammals	Risk ratio (PEC/LD_{50})	<0.01	0.01–0.1	0.1	EPPO
Bees	Risk ratio (recommended dose rate/LD_{50})	<50	50–500	>500	PRG/EPPO*
Other terrestrial arthropods	Acute toxicity (%) at recommended dose rate	<50%	50–99%	>99%	IOBC†

B. Field data (well-conducted field trials and control operations)

Group	Parameter	Risk class			References
		low (L)	Medium (M)	High (H)	
Fish	Evidence of mortality	None	Incidental	Massive	PRG
Aquatic arthropods	Population reduction	<50%	50–90%	>90%	PRG
Reptiles, birds, mammals	Evidence of mortality	None	Incidental	Massive	PRG
Bees	Evidence of mortality	Not significant	Incidental	Massive	EPPO
Other terrestrial arthropods	Population reduction	<25%	25–75%	>75%	IOBC

As a result of a greater error associated with population estimates of terrestrial arthropods, the lower limits of the different risk classes are lower than for aquatic arthropods.
PEC, predicted environmental concentration after treatment at the recommended dose rate; LC_{50}, median lethal concentration; LD_{50}, median lethal dose; FAO/Locustox, FAO Locustox project in Senegal (Everts et al., 1997, 1998); EPPO, European and Mediterranean Plant Protection Organization (EPPO, 2003a); PRG, Pesticide Referee Group.
* EPPO (2003b).
† International Organization for Biological and Integrated Control of Noxious Animals and Plants (Hassan, 1994).

In the UK there is a Wildlife Incident Investigation Scheme, which provides information on hazards to wildlife and companion animals (usually cats and dogs) and beneficial invertebrates (e.g. bees) from pesticides; and is responsible for identifying and penalising those who deliberately or recklessly misuse and abuse pesticides (Fig. 6.19). The scheme also includes suspect baits, where it is thought that pesticides, such as carbofuran have been inappropriately applied or used, and spillages of pesticides where this poses a risk to wildlife or companion animals.

The use of rodenticides remains a problem as rats or other poisoned animals may be outside buildings and get eaten by birds of prey. Thus certain

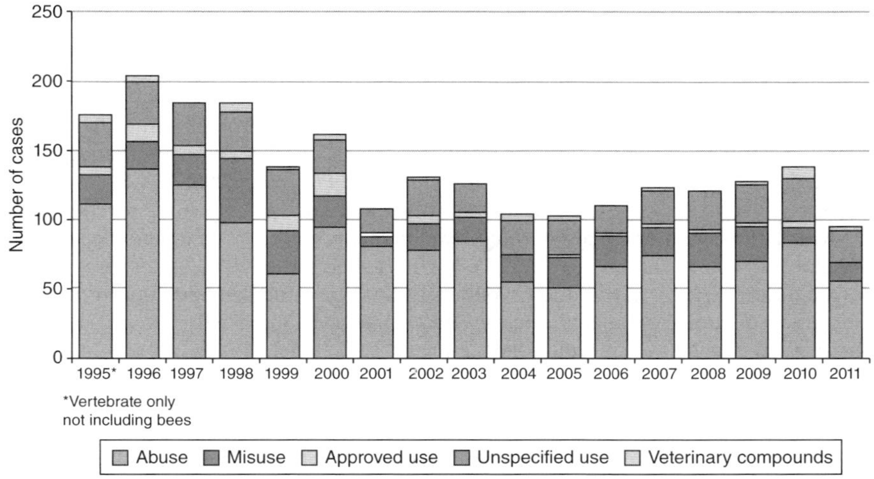

Fig. 6.19 Pesticide poisoning incidents investigated by the Wildlife Incident Investigation Scheme (WIIS) in the UK.

rodenticides may only be used indoors, although where resistance occurs limited use outdoors may be possible following strict procedures.

This chapter has shown that the authorities have responded to problems encountered when pesticides cause unacceptable adverse effects in the environment. The requirements for registration now require far more environmental data and great care is taken to look at all aspects of potential problems. Inevitably some chemicals have been registered as subsequent adverse effects were not foreseen DDT is a good example of a very effective insecticide of low mammalian toxicity, but its effects on birds and the food chain had not been realised when it was initially promoted. As our knowledge base continues to expand, conditions for registration have become more precise.

References

Anon (2001) *Groundwater Protection Code: Use and Disposal of Sheep Dip Compounds*. DEFRA, London.

Anon (2004) Report of the Locust Pesticide Referee Group. 9th Meeting. FAO, Rome. Available at: http://www.fao.org/ag/locusts-CCA/common/ecg/1013/en/PRG_9_2004_Eng.pdf (accessed on 7 May 2015).

Anon (2009) *Agri-Environment Schemes in England 2009: A Review of Results and Effectiveness*. Natural England Publications, London.

Anon (2014a) *The National Pollinator Strategy: For Bees and Other Pollinators in England*. DEFRA, London.

Anon (2014b) *Supporting Document to the National Pollinator Strategy: for Bees and Other Pollinators in England*. DEFRA, London.

Arias-Este, M., Lopez-Periago, E., Martinez-Carballo, E., Simal-Gandara, J., Mejuto, J-C. and Garciia-Rio, L. (2008) The mobility and degradation of pesticides in soils and the pollution of groundwater resources. *Agriculture, Ecosystems and Environment* **123**, 247–260.

Armbrust, K.L. and Peeler, H.B. (2002) Effects of formulation on the run-off of imidacloprid from turf. *Pest Management Science* **58**, 702–706.

Arnold, G.L., Luckenbach, M.W. and Unger, M.A. (2004) Runoff from tomato cultivation in the estuarine environment: biological effects of farm management practices. *Journal of Experimental Marine Biology and Ecology* **298**, 323–346.

Ashauer, R. and Brown, C.D. (2013) Highly time-variable exposure to chemicals: toward an assessment strategy. *Integrated Environmental Assessment and Management* **9**, 27–33.

Barraclough, D., Kearney, T. and Croxford, A. (2005) Bound residues: environmental solution or future problem? *Environmental Pollution* **133**, 85–90.

Bartuszevige, A.M., Capparella, A.P., Harper, R.G., Frick, J.A., Criley, B., Doty, K. and Erhart, E. (2002) Organochlorine pesticide contamination in grassland-nesting passerines that breed in North America. *Environmental Pollution* **117**, 225–232.

Berenzen, N., Kumke, T., Schultz, H.K. and Schultz, R. (2005a) Macroinvertebrate community structure in agricultural streams: impact of runoff-related pesticide contamination. *Ecotoxicology and Environmental Safety* **70**, 37–46.

Berenzen, N., Lentzen-Godding, A., Probst, M., Schultz, H., Schultz, R. and Liess, M. (2005b) A comparison of predicted and measured levels of runoff-related pesticide concentrations in small lowland streams on a landscape level. *Chemosphere* **58**, 683–691.

Blanchoud, H., Farrugia, F. and Mouchel, J.M. (2004) Pesticide uses and transfer in urbanised catchments. *Chemosphere* **55**, 905–913.

Brown, C.D.,Bellamy, P.H. and Dubus, I.G. (2002) Prediction of pesticide concentrations found in rivers in the UK. *Pest Management Science* **58**, 363–373.

Brown, C.D., Hart, A., Lewis, K.A. and Dubus, I.G. (2003) p-EMA (I): simulating the environmental fate of pesticides for a farm-level risk assessment system. *Agronomie* **23**, 67–74.

Brown, C.D., Dubus, I.G., Fogg, P., Spirlet, M. and Gustin, C, (2004a) Exposure to sulfosulfuron in agricultural drainage ditches: field monitoring and scenario-based modelling. *Pest Management Science* **60**, 765–776.

Brown, R.B., Carter, M.H. and Stephenson, G. R. (2004b) Buffer zone and windbreak effects on spray deposition in a simulated wetland. *Pest Management Science* **60**, 1085–1090.

Butler-Ellis, C. and Bradley, A. (2002) The influence of formulation on spray drift. *Aspects of Applied Biology* **66**, 251–258.

Byron, N. and Hamey, P. (2008) Setting unsprayed buffer zones in the UK. *Aspects of Applied Biology* **84**, 123–129.

Caguan, A.G. (1995) The impact of pesticides on ricefield vertebrates with emphasis on fish. In Pingali, P.L. and Roger, P.A. (eds.) *Impact of Pesticides on Farmer Health and the Rice Environment*. Kluwer, Dordrecht, pp. 203–248.

Cardinali, A., Otto, S. and Zanin, G. (2013) Herbicides runoff in vegetative filter strips: evaluation and validation of a recent rainfall return period model. *International Journal of Environment Analytical Chemistry* **93**, 1628–1637.

Carluer, N., Tournebize, J., Gouy, V., Margoum, C., Vincent, B. and Gril, J.J. (2011) Role of buffer zones in controlling pesticide fluxes to surface waters. *Procedia Environmental Science* **9**, 21–26.

Collins, K.L., Boatman, N.D., Wilcox, A. Holland, J.M. and Chaney, K. (2002) Influence of beetle banks on cereal aphid predation in winter wheat. *Agriculture, Ecosystems and Environment* **93**, 337–350.

Cooke, A.S. (1973) Shell thinning in avian eggs by environmental pollutants. *Environmental Pollution* **4**, 85–152.
Crane, M. Whitehouse, P., Comber, S., Watts, C., Giddings, J. Moore, D.R.J. and Grist, E. (2003) Evaluation of probabilistic risk assessment in the UK: chlorpyrifos use on top fruit. *Pest Management Science*, **59**, 512–526.
Cross, P., and Edwards-Jones, G. (2006) Variation in pesticide hazard from arable crop production in Great Britain from 1992 to 2002: pesticide risk indices and policy analysis. *Crop Protection* **25**, 1101–1108.
Cross, J.V., Murray, R.A., Walklate, P.J. and Richardson, G.M. (2004) Pesticide dose Adjustment to the Crop Environment (PACE): efficacy evaluations in UK apple orchards 2002–2003. *Aspects of Applied Biology* **71**, 287–294.
Cunha J.P.,Chueca, P., Garcerá, C. and Moltó, E. (2012) Risk assessment of pesticide spray drift from citrus applications with air-blast sprayers in Spain. *Crop Protection* **42**, 116–123.
Cutler, G.C., Scott-Dupree, C.D., Sultan, M., McFarlane, A.D. and Brewer, L. (2014) A large-scale field study examining effects of exposure to clothianidin seed-treated canola on honey bee colony health, development, and overwintering success. PeerJ, 2e652, doi:10.7717.652.
Dabrowski, J.M., Shadung, J.M. and Wepener, V. (2014) Prioritizing agricultural pesticides used in South Africa based on their environmental mobility and potential human health effects. *Environment International* **62**, 31–40.
Davis, B.N., Brown, M.J. and Frost, A.J. (1993) Selection of receptors for measuring spray drift deposition and comparison with bioassays with special reference to the shelter effect of hedges. *Brighton Crop Protection Conference – Weeds*, BCPC, Farnham, UK, pp. 139–144.
Davis, B.N., Frost, A.J. and Yates, T.J. (1994a) Bioassays of insecticide drift from air assisted sprayers in an apple orchard. *Journal of Horticultural Science* **69**, 703–708.
Davis, B.N., Brown, M.J., Frost, A.J., Yates, T.J. and Plant R.A. (1994b) The effects of hedges on spray deposition and on the biological impact of pesticide spray drift. *Ecotoxicology and Environmental Safety* **27**, 281–293.
De Snoo, G.R. (1999) Unsprayed field margins: effects on environment, biodiversity and agricultural practice. *Landscape and Urban Planning* **46**, 151–160.
De Snoo, G.R and de Witt, P.J. (1998) Buffer zones for reducing pesticide drift to ditches and risks to aquatic organisms. *Ecotoxicology and Environmental Safety* **41**, 112–118.
De Snoo, G.R and van der Poll R.J. (1999) Effect of herbicide drift on adjacent boundary vegetation. *Agriculture, Ecosystems and Environment* **73**, 1–6.
Dewar, A.M. (1994) The virus yellows warning system – an integrated pest management system for beet in the UK. In: Leather, S.R., Mills, N.J. and Walters, K.F.A. (eds.) *Individuals, Populations and Patterns in Ecology*. Atheneum Press, Newcastle on Tyne, pp. 172–185.
Douthwaite, R.J. and Tingle, C.C.D. (1994) *DDT in the Tropics: The Impact on Wildlife in Zimbabwe of Ground-Spraying for Tsetse Fly Control*. Natural Resources Institute, Chatham.
Elliott, J.G. and Wilson, B.J. (1983) *The Influence of Weather on the Efficiency and Safety of Pesticide Application: The Drift of Herbicides*. Occasional Publication No. 3. British Crop Protection Council, Farnham, UK.
EPA (1992) *Developing a Work Scope for Ecological Assessments*. Eco Update; Intermittent Bulletin, Vol. 1, No. 4. U.S. Environmental Protection Agency, Washington, DC, USA.
EPPO/Council of Europe (2003a) Environmental assessment scheme of plant protection products – Chapter 10: Honeybees. *OEPP/EPPO Bulletin* **33**, 141–145.

EPPO/Council of Europe (2003b) Environmental assessment scheme of plant protection products – Chapter 11: Terrestial vertebrates. *OEPP/EPPO Bulletin* **33**, 211–238.

EU Commission (2010) Commission Directive 2010/21/EU of 12th March amending Annex 1 to Council Directive 91/414/EEC as regards the specific provisions relating to clothianidin, thiamethoxam, fipronil and imidacloprid. Official Journal of the European Union L65/27.

Everts, J.W., Mbaye, D. and Barry, O. (eds.) (1997) *Environmental Side Effects of Locust and Grasshopper Control* Vol. **1**. FAO:GCP/SEN/053/NET. Dakar, Rome.

Everts, J.W., Mbaye, D., Barry, O. and Mullie, W. (eds.) (1998) *Environmental Side Effects of Locust and Grasshopper Control*, Vols. **2 and 3**. FAO:GCP/SEN/053/NET. Dakar, Rome.

Farooq, M., Balachandar, R., Wulfson, D. and Wolf, T.M. (2001) Agricultural sprays in cross-flow and drift. *Journal of Agricultural and Engineering Research* **78**, 347–358.

Foque, D., Devarrewaere, W., Verboven, P. and Nuyttens, D. (2014a) Characterisation of different pneumatic sowing machines. *Aspects of Applied Biology* **122**, 77–84.

Foque, D., Devarrewaere, W., Verboven, P. and Nuyttens, D. (2014b) Physical and chemical characteristics of abraded seed coating particles. . *Aspects of Applied Biology* **122**, 85–94.

Fox, P.J. and Matthiessen, P. (1982) Acute toxicity to fish of low-dose aerosol applications in relation to control tsetse fly in the Okavango Delta, Botswana. *Environmental Pollution Series A* **27**, 129–142.

Ganzelmeier, R. (2005) GIS-based application of plant protection products – examples from research and application. Paper presented at the Conference on Environmentally Friendly Application Techniques. Warsaw, Poland, October 2004. Annual Review of Agricultural Engineering 4, 245–255.

Ganzelmeier, R. and Rautmann, D (2000) Drift, drift reducing sprayers and sprayer testing. *Aspects of Applied Biology* **57**, 1–10.

Ganzelmeier, R., Rautmann, D., Spangenberg, R., Streloke, M., Herrmann, M., Wenzelburger, H.J. and Walter, H.F. (1995) Studies on the spray drift of plant protection products. 111pp, BBA, Berlin.

Geiger, F., Bengtsson, J., Berendse, F.F.B.N., Weisser, W.W., Emmerson, M., Morales, M.B., Ceryngier, P., Liira, J., Tscharntke, T., Winqvist, C., Eggers, S., Bommarco, R., Part, T., Bretagnolle, V., Plantegenest, M., Clement, L.W., Dennis, C., Palmer, C., Onate, J.J., Guerrero, I., Hawro, V., Aavik, T., Thies, C., Flohre, A., Haenke, S., Fischer, C., Goedhart, P.W. and Inchausti, P. (2010) Persistent negative effects of pesticides on biodiversity and biological control potential on European farmland. *Basic and Applied Ecology* **11**, 97–105.

Gilbert, A.J. (2000) Local environmental risk assessment for pesticides. *Aspects of Applied Biology* **57**, 83–90.

Gove, B. (2004) The impact of pesticide spray drift and fertilizer over-spread on the ground flora of ancient woodland. PhD Thesis. University of London, London.

Gove, B., Ghazoury, J., Power, S. and Buckley, P (2004) The Impacts of Pesticide Spray Drift and Fertiliser Over-Spread on the Ground Flora of Ancient Woodland. English Nature Research Reports. English Nature, Peterborough, p. 614.

Grant, I.F. and Tingle C.D. (2002) *Ecological Monitoring Methods*. Handbook Natural Resources Institute, University of Greenwich, Greenwich, 266pp.

Hassan, S.A. (1994) Activities of the IOBC/WPRS working group 'Pesticides and Beneficial Organisms'. *IOBC/WPRS Bulletin* **17**, 1–5.

Rasmussen, L.H., Kroghsbo, S., Frisvad, J.C. and Hansen, H.C. (2003) Occurrence of the carcinogenic Bracken constituent ptaquiloside in fronds, topsoils and organic soil layers in Denmark *Chemosphere* **51**, 117–127.

Hart, A., Brown, C.D., Lewis, K.A. and Tzilivakis, J. (2003) p-EMA (II): evaluating ecological risks of pesticides for farm-level risk assessment system. *Agronomie* **23**, 75–84.

Herbst, A. (2003) Pesticide formulation and drift potential. *The BCPC International Congress*, BCPC, Farnham, UK, pp. 255–260.

Hewitt, A.J. (2000) Spray drift modelling, labelling and management in the US. *Aspects of Applied Biology* **57**, 11–19.

Hickey, J.J. and Anderson, D.W. (1968) Chlorinated hydrocarbons and eggshell changes in raptorial and fish-eating birds *Science* **162**, 271–273.

Holland, J.M. Oaten, H., Moreby, S., Birkett, T., Simper, J.,Southway, S. and Smith, B.M. (2012) Agri-environment scheme enhancing ecosystem services: a demonstration of improved biological control in cereal crops. *Agriculture, Ecosystems and Environment* **155** 147–152.

Irina Herzon, I., Ekroos, J., Rintala, J., Tiainen, J., Seimola, T. and Vepsäläinen, V. (2011) Importance of set-aside for breeding birds of open farmland in Finland. *Agriculture, Ecosystems and Environment* **143**, 3–7.

Jacquet, F., Butault, J-P. and Guichard, L. (2010) An economic analysis of the possibility of reducing pesticides in French field crops. European Association of Agricultural Economists 120th Seminar, 2–4 September 2010, Chania, Crete.

Koch, H., Weisser, P. (2004) A proposal for a higher tier investigation of pesticide drift exposure to non-target organisms (NTO) in field trials. *Nachrichtenblatt des deutschen pflanzenschutzdienstes* **56S**, 180–183.

Koch, H., Strub, O. and Weisser, P. (2004a) The patchiness of pesticide drift deposition patterns in plant canopies. *Nachrichtenblatt des deutschen pflanzenschutzdienstes* **56S**, 25–29.

Koch, H., Weisser, P. and Strub, O. (2004b) Comparison of dose response of pesticide spray deposits versus drift deposits. *Nachrichtenblatt des deutschen pflanzenschutzdienstes* **56S**, 30–34.

Kookana, R.S. and Aylmore, L.A.G. (1994) Estimating the pollution potential of pesticides to groundwater. *Australian Journal of Soil Research* **32**, 1141–1155.

Kookana, R.S., Correll, R.L. and Miller, R.R.B. (2005) Pesticide impact rating index – a pesticide risk indicator for water quality. *Water, Air, and Soil Pollution* **5**, 45–65.

Kuussaaria, M., Hyvönen, T. and Härmäc, O. (2011) Pollinator insects benefit from rotational fallows. *Agriculture, Ecosystems and Environment* **143**, 28–36.

Lane, A.G. and Butler-Ellis, C. (2003) Assessment of environmental concentrations of pesticide from spray drift. *The BCPC International Congress*, BCPC, Farnham, UK, pp. 501–506.

Lewis, K.A., Brown, C.D., Hart, A. and Tzilivakis, J. (2003) p-EMA (III): overview and application of a software system designed to assess environmental risk of agricultural pesticides. *Agronomie* **23**, 85–96.

Maeda, T. (2005) Bird use of rice field strips of varying width in the Kanto Plain of central Japan. *Agriculture, Ecosystems and Environment* **105**, 347–351.

Margni, M., Rossier, D., Crettaz, P. and Jolliet, O. (2002) Life cycle impact assessment of pesticides on human health and ecosystems. *Agriculture, Ecosystems and Environment* **93**, 379–392.

Marzaro, M., Vivan, L., Targa, A., Mazzon, L., Mori, N., Greatti, M., Toffolo, E.P., Di Bernardo, A., Giorio, C., Marton, D., Tapparo, A. and Girolami, V. (2011) Lethal aerial powdering of honey bees with neonicotinoids from fragments of maize seed coat. *Bulletin of Insectology* **64**, 119–126.

Metcalf, R.L. (1984) An increasing public concern. *EPA Journal* **10**, 30–31. Also in Pimental, D. and Lehman, H (eds.) (1993) *The Pesticide Question*. Chapman and Hall, New York.

Miller D.R. and Stoughton, T.E. (2000) Response of spray drift from aerial applications at forest edge to atmospheric stability. *Agricultural and Forestry Meteorology* **100**, 49–58.

Morris, A.J., Wilson, J.D., Whittingham, M.J. and Bradbury, R. (2005) Indirect effects of pesticides on breeding yellowammer (*Emberiza citrinella*). *Agriculture, Ecosystems and Environment* **106**, 1–16.

Muller, K., Deurer, M., Hartmann, H., Bach, M., Spiteller, M. and Frede, H-G. (2003) Hydrological characterisation of pesticide loads using hydrograph separation at different scales in a German catchment. *Journal of Hydrology* **273**, 1–17.

Nakano, Y., Miyazaki, A., Yoshida, T., Ono, K. and Inoue, T. (2004) A study on pesticide runoff in the Kozakura River, Japan. *Water Research* **38**, 3017–3022.

Neumann, M., Schulz, R., Schafer, K., Muller, W., Mannheller, W. and Liess, M. (2002) The significance of entry routes as point and non-point source of pesticides in small streams. *Water Research* **36**, 835–842.

Newton, I. (1998) *Population Limitation in Birds*. Academic Press, San Diego, CA.

Nowell, L.H., Norman, J.E., Moran, P.W., Martin, J.D. and Stone, W.W. (2014) Pesticide Toxicity Index—a tool for assessing potential toxicity of pesticide mixtures to freshwater aquatic organisms. *Science of the Total Environment* **476–477**, 144–157.

Nuyttens, D., Devarrewaere, W., Verboven, P. and Foque, D. (2013) Pesticide-laden dust emission and drift from treated seeds during seed drilling: a review. *Pest Management Science* **69**, 564–575.

Ogada, D.L.(2014) The power of poison: pesticide poisoning of Africa's Wildlife. *Annals of the New York Academy of Sciences* **1322**, 1–20.

Padovani, L. Trevisan, M. and Capri, E. (2004) A calculation procedure to assess potential environmental risk of pesticides at the farm level. *Ecological Indicators* **4**, 111–123.

Parish, D.M.B. and Sotherton, N.W. (2004) Game crops as summer habitat for farmland songbirds in Scotland. *Agriculture Ecosystems and Environment* **104**, 429–438.

Parkin, C.S., Walklate, P.J. and Nicholls, J.W. (2003) Effect of drop evaporation on spray drift and buffer zone risk assessments. *The BCPC International Congress*, BCPC, Farnham, UK, pp. 261–266.

Pistorius, J. Bischoff, G., Heimbach, U. and Stahler, M. (2009) Bee poisoning incidents in Germany in Spring 2008 caused by abrasion of active substances from treated seeds during sowing of maize. *Julius Kuhn Archives* **423**, 118–125.

Planas, S., Solanelles, F. and Fillat, A. (2002) Assessment of recycling tunnel sprayers in Mediterranean vineyards and apple orchards. *Biosystems Engineering* **82**, 45–52.

Potts, G.R. (1986) *The Partridge: Pesticides, Predation and Conservation*. Collins, London.

Radcliffe, D.A. (1967) Decrease in eggshell weight in certain birds of prey. *Nature* **215**, 208–210.

Radcliffe, D.A. (1970) Changes attributable to pesticides in egg breakage frequency and egg shell thickness in some British birds. *Journal of Applied Ecology* **7**, 67–115.

Ramwell, C.T., Heather, A.I.J. and Shepherd, A.J. (2002) Herbicide loss following application to a roadside. *Pest Management Science* **58**, 695–701.

Ramwell, C.T., Heather, A.I.J. and Shepherd, A.J. (2004) Herbicide loss following application to a railway. *Pest Management Science* **60**, 556–564.

Reichenberger, S., Bach, M., Skitschak, A. and Frede, H-G. (2007) Mitigation strategies to reduce pesticide inputs into ground- and surface water and their effectiveness: a review. *Science of the Total Environment* **384**(2007), 1–35.

Reus, J., Leendertse, P., Bockstaller, C., Fomsgaard, I., Gutsche, V., Lewis, K., Nilsson, C., Pussemier, L., Trevisa, M., van der Werf, H., Alfarroba, F., Blumel, S., Isart, J.,

McGath, D. and Seppala, T. (2002) Comparison and evaluation of eight pesticide environmental risk indicators developed in Europe and recommendations for future use. *Agriculture, Ecosystems and Environment* **90**, 177–187.

Robinson, R.C., Parsons, R.G., Barbe, G., Patel, P.T. and Murphy, S. (2000) Drift control and buffer zones for helicopter spraying of bracken (*Pteridium aquilinum*) *Agriculture, Ecosystems and Environment* **79**, 215–231.

Ropke, B., Bach, M and Frede, H-G. (2004) DRIPS – a DSS for estimating the input quantity of pesticides for German river basins. *Environmental Modelling and Software* **19**, 1021–1028.

Sanchez-Bayo, F.,Ward, R. and Beasley, H. (1999) A new technique to measure bird's dietary exposure to pesticides. *Analytica Chimica Acta* **399**, 173–183.

Sanchez-Bayo, F., Baskaran, S. and Kennedy, I.R. (2002) Ecological relative risk (EcoRR): another approach for risk assessment of pesticides in agriculture. *Agriculture, Ecosystems and Environment* **91**, 37–57.

Sankararamakrishnan, N., Sharma, A.K. and Sanghi, R. (2004) Organochlorine and organophosphorous pesticide residues in ground water and surface waters of Kanpur, Uttar Pradesh, India. *Environmental International* **31**, 113–120.

Schultz, R. (2001) Rainfall-induced sediment and pesticide input from orchards into the Lourens River, Western Cape South Africa: importance of a single event. *Water Research* **35**, 1869–1876.

Sidahmed, M.M., Awadalla, H.H. and Haidar, M.A. (2004) Symmetrical multi-foil shields for reducing spray drift. *Biosystems Engineering* **88**, 305–312.

Skark, C., Zullei-Seibert, N., Willme, U., Gatzemann, U. and Schlett, C. (2004) Contribution of non-agricultural pesticides to pesticide load in surface water. *Pest Management Science* **60**, 525–530.

Tariq, M.I., Afzal, S. and Hussain, I. (2003) Pesticides in shallow groundwater of Bahawalnagar, Muzafargarth, D.G. Khan and Rajan Pur districts of Punjab, Pakistan. *Environmental International*. **30**, 471–79.

Thomas, C.F.G. and Marshall, E.J.P. (1999) Arthropod abundance and diversity in different vegetated margins of arable fields. *Agriculture, Ecosystems and Environment* **72**, 131–144.

Toan, P.V., Sebesvari, Z., Bläsing, M., Rosendahl, I. and Renaud, F.G. (2013) Pesticide management and their residues in sediments and surface and drinking water in the Mekong Delta, Vietnam. *Science of the Total Environment* **452–453**, 28–39.

Toccalino, P.L., Gilliom, R.J., Lindsey, B.D. and Rupert, M.G. (2014) Pesticides in groundwater of the United States: Decadel-Scale Changes, 1993–2011. Groundwater, doi:10.1111/gwat.12176.

Tournebize, J., Vincent, B., Chaumont, C., Gramaglia, C., Margoum, C., Molle, P., Carluer, N. and Gril, J.J. (2011) Ecological services of artificial wetland for pesticide mitigation. Socio-technical adaptation for watershed management through TRUSTEA project feedback. *Procedia Environmental Sciences* **9**, 183–190.

Tournebize, J. Passeport, E., Chaumont, C., Fesneau, C., Guenne, A. and Vincent, B. (2013) Pesticide de-contamination of surface waters as a wetland ecosystem service in agricultural landscapes. *Ecological Engineering* **56**, 51–59.

Tscharntke, T., Klein, A.M., Kruess, A., Steffan-Dewenter, I., and Thies, C. (2005) Landscape perspectives on agricultural intensification and biodiversity—ecosystem service management. *Ecological Letters* **8**, 857–874.

Ucar, T., Hall, F.R., Tew, J.E. and Hacker, J.K. (2003) Wind tunnel studies on spray deposition on leaves of tree species used in windbreaks and exposure to bees. *Pest Management Science* **59**, 358–364.

Van Wyk, E., Bouwman, H., van der Bank, H., Verdoorn, G.H. and Hofmann, D. (2001) Persistent organochlorine pesticides detected in blood and tissue samples of vultures from different localities in South Africa. *Comparative Biochemistry and Physiology Part C* **129**, 243–264.

Walters, K.F.A. (2013) Data, data everywhere but we don't know what to think? Neonicotinoid insecticides and pollinators. *Outlooks on Pest Management* **24**, 151–155.

Wenneker, M. and Van de Zande, J.C. (2008) Spray drift reducing effects of natural windbreaks in orchard spraying. *Aspects of Applied Biology* **84**, 25–32.

Woltza, J.M., Isaacs, R. and Landisa, D.A. (2012) Landscape structure and habitat management differentially influence insect natural enemies in an agricultural landscape. *Agriculture, Ecosystems and Environment* **152**, 40–49.

Woodcock, B.A., Westbury, D.B., Potts, S.G., Harris, S.J. and Brown, V.K. (2005) Establishing field margins to promote beetle conservation in arable farms. *Agriculture, Ecosystems and Environment* **107**, 255–266.

7 Residues in food

Some members of the public are against all use of pesticides, hence the promotion of 'organic' food. They forget that since man changed from a hunter gatherer to being a farmer, plants have been selected which provide us with carbohydrates, notably maize, rice and wheat (Fig. 7.1), and proteins from the peas and beans, the latter also helping to improve the soil due to the their ability to fix atmospheric nitrogen. Over the centuries, selection of higher yielding crops developed into more scientific plant breeding looking for resistance to major pests and diseases. Outside these main crops we like to have beverages, fruit and vegetables which have their own unique tastes as they contain unusual chemicals. Coffee is an extremely popular drink, but it contains caffeine, which is a crystalline, xanthine alkaloid, that is a stimulant drug. The presence of this in the coffee plant is a natural pesticide to protect it from certain insects that would otherwise feed on it. In humans, caffeine acts on the central nervous system and keeps us alert. The amount of caffeine in a single cup is quite small, so our digestive system can cope with it, but if an adult consumed over 10 g of caffeine rather than 500 mg, it would have a toxic effect. The median lethal dose (LD_{50}) of caffeine is 192 mg/kg which is approximately the equivalent of 80–100 cups of coffee. There are other drinks and vegetables that are safe despite containing very small quantities of a natural pesticide as the human body can metabolise the substance and suffer no ill effects.

When the modern pesticides were marketed, the regulatory authorities soon established the use of maximum residue levels (MRLs) that are derived from field trials carried out according to good agricultural practice (GAP), including observance of a pre-harvest interval (PHI) to determine what is the highest legally permitted residue that could be present in a crop. An MRL is intended to be a legally enforceable limit to check whether farmers do follow GAP. It is not based on the acceptable daily intake (ADI) of a pesticide residue. An MRL is usually derived from data obtained from 8 to 16 field trials (Hyder and Travis, 2003), but as the agrochemical companies concentrate on major crops, in many situations the MRL is set at the limit of detection of the pesticide and its metabolites. MRLs are a mechanism for

Pesticides: Health, Safety and the Environment, Second Edition. G.A. Matthews.

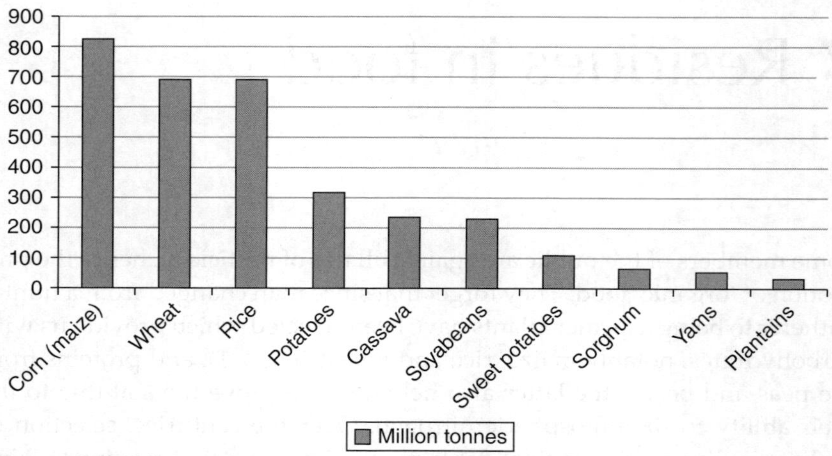

Fig. 7.1 Global production of the 10 major food crops in 2008.

Table 7.1 Examples of estimates of the dietary intake of pesticide residues for commodities based on the 90th percentile for residues (European Commission, 1998)

Pesticide	Food	90th percentile (mg/kg)	ADI (μg/kg bw)	Average consumption (g/day)	Intake (μg/person/day)	Intake (μg/kg bw/day)	% ADI
Chlorpyrifos	Apples	<0.03	10	40	1.2	0.02	0.2
Methamidophos	Lettuce	<0.04	4	22.5	0.9	0.015	0.38
Iprodione	Lettuce	<1	60	22.5	22.5	0.375	0.63
Procymidone	Lettuce	<0.03	100	22.5	0.675	0.011	0.01
Chlorothanonil	Lettuce	0.02	30	22.5	0.45	0.008	0.03
Maneb	Lettuce	<9	30	22.5	203.5	3.375	11.25

From European Commission (1998).

regulating trade in produce and are generally set at a level many times lower than a level, which would be expected to have an adverse effect on human health. Consuming foodstuffs with residues in excess of an MRL does not, therefore, necessarily constitute a risk to consumer health. Some examples of the relationship between the pesticide residues and the ADI are shown in Table 7.1.

Farmers have to follow advice and ensure that the pre-harvest interval (PHI) is observed to minimise the residue that is within the harvested crop. This is the period between the last application and harvesting. One study on peaches indicated a PHI of 27 days was needed when spraying chlorpyrifos (Chatzicharisis et al., 2012). When a crop has to be treated on several occasions and is harvested over a long period, it is more difficult to check that the interval is fully effective and ensures residues are within the allowed limits. In these situations, pesticides with a known short persistence are preferable.

This contrasts with the situation when organochlorine insecticides were used, as they could persist on the harvested produce over a long period. The persistent organochlorine pesticides (POPs) are no longer permitted to be used.

Only a fraction of the pesticide applied on a crop is actually retained on plants, as applying a foliar spray is still a very inefficient process, with some spray lost to the ground or downwind as spray drift, although improved technology can reduce the drift. The amount that is retained on the crop depends on many factors. These include the formulation of pesticide used, the volume of spray applied, type of equipment used, such as with air assistance and the quality of the spray, for example the droplet sizes. The physical characteristics of the foliage, especially the leaf surface will also affect whether droplets bounce off leaves or are retained. When the volume of spray liquid applied was high (>500 l/ha), in some cases only about 20% of the pesticide was deposited on the crop, with most of the chemical wasted on the soil (Fig. 7.2). The trend has been to reduce the spray volume and many sprays are now applied at less than 150 l/ha, so increasing spray efficiency, through reduced run-off and better retention by the crop, although as mentioned previously, if droplet size is too large, more pesticide is likely to be deposited on the soil.

The droplets, which remain on the surface of the plant, allow some of the active ingredient to penetrate through the leaf cuticle, with the remainder drying to form a surface deposit. From this deposit, the active ingredient will either stay on the surface to control pests that walk or land on the outside of the plant, penetrate through the outer surface of the plant or be lost from the surface. The latter may be caused by rain, especially if it occurs with an hour or so of the pesticide being applied, or it may volatilise from the surface or be abraded physically by other surfaces. Further losses occur as the active ingredient is degraded either within the plant or on the surface, for example by ultra-violet light. The concern in terms of residues in crops is the amount of active ingredient that remains on or within the harvested product.

A check on the residue level in a crop is made by sampling crops and produce on sale to determine whether there is any detectable residue in them (Figs. 7.3, 7.4 and 7.5). Official data in the UK obtained in these samples are assessed by the Expert Committee on Pesticide Residues in Food (PRiF) and is published. The results indicate how many samples have no detectable residue and from those with a residue, whether the amount of the active ingredient exceeds the MRL at or shortly after harvest. Residue data are also obtained also from processed food, such as bread and pasta, or on commodities not grown in the country, but are imported, for example on bananas. Similar studies are conducted throughout Europe and some may be carried out by individual companies.

Within the EU, apples, head cabbage, leek, lettuce, milk, peaches, pears, rye or oats, strawberries, swine meat and tomatoes were the commodities

(a)

(b)

Fig. 7.2 Excessive spray deposits on (a) lettuce and (b) tomato crops (Photographs by Graham Matthews).

analysed in the framework of the 2010 EU-coordinated control programme. A total number of 12,168 samples were analysed in 2010, of which 50.7% were free from measurable pesticide residues; 47.7% of the samples had measurable residues below or at the MRL, while only 1.6% of the samples had residues that exceeded the MRL. EFSA concluded that the long-term

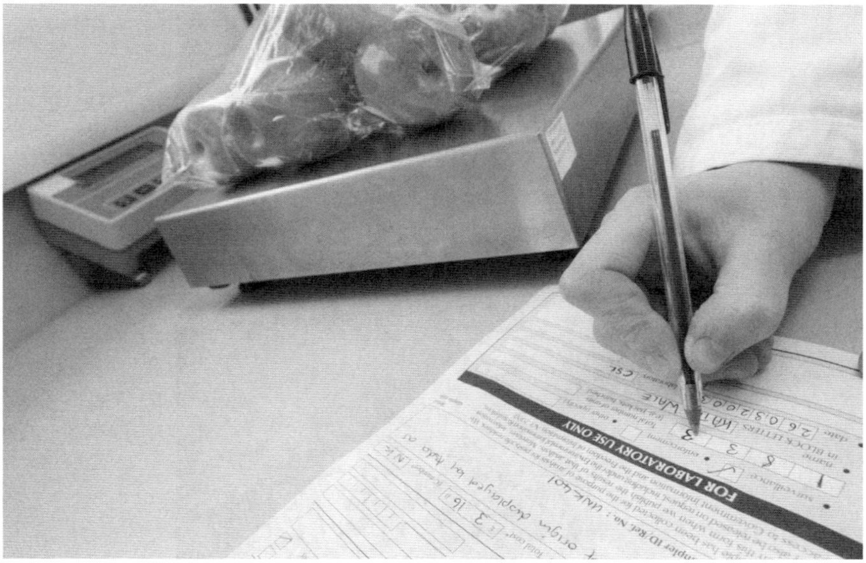

Fig. 7.3 Recording details of samples for residue analysis (Central Science Laboratory).

Fig. 7.4 Processing samples for residue analysis (Central Science Laboratory).

exposure of consumers to these commodities did not raise health concerns, but in assessing the short-term exposure, there was a risk associated with 79 samples if the pertinent food was consumed in high amounts. It was also reported that national authorities within the EU analysed more than 77,000

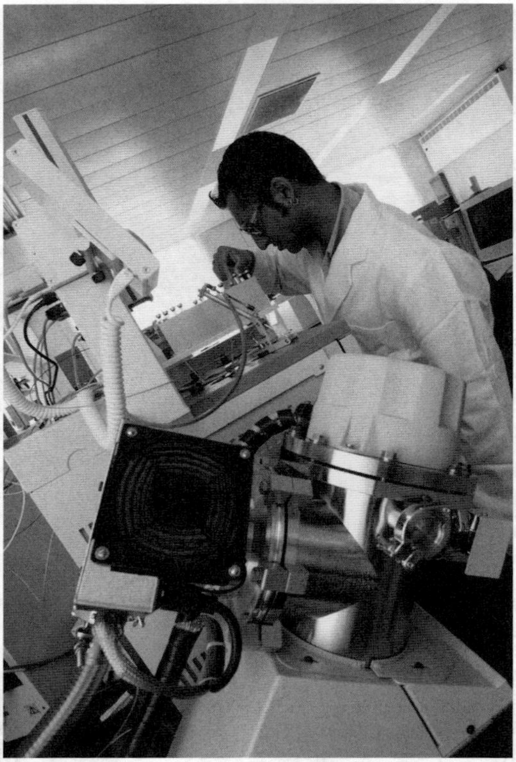

Fig. 7.5 Analysis of sample extracts by gas chromatography (Central Science Laboratory).

samples of approximately 500 different types of food for pesticide residues during 2010 (EFSA, 2013).

Data from a selection of crops show the number of samples analysed for certain crops in recent years in the UK (Table 7.2). In 2011, only 8 (1.09%) of the 737 samples of UK fruit and vegetables had residues above the MRL, whereas 5% of 1237 samples of non-UK food and drink tested, contained residues above the MRL (Fig. 7.6). Despite the low proportion of samples, media coverage always paints a more alarming picture, such as 93% of the (non-organic) oranges you buy have residues, without any indication of which residues were present or that the presence of small residues is not a health concern.

Samples of Polish apples from 2005 to 2013 (Lozowicka, 2015) detected 34 pesticides in 66.5% of the samples with multiple residues in 35% of samples. The chronic and acute exposures were assessed using a deterministic model that was based on the average and high concentrations of residues. In certain cases, the total dietary pesticide intake calculated from the residue levels observed in apples exceeded the toxicological criteria. In the cumulative chronic exposure, among the 17 groups of compounds studied, organophosphate

Table 7.2 Pesticide residues detected between 1991 and 2002 for selected crops in the UK as reported by the Pesticides Residues Committee (Adapted from Foster et al., 2003)

Crop	No. of samples tested	% with residues	%>MRL	No. of pesticides found
Apple	396	44	0	25
Banana	181	65	2.8	7
Carrot	369	64	0.8	12
Celery	276	66	4.0	30
Grapes	382	44	2.1	46
Lettuce	803	58	3.7	37
Mushroom	255	11	0.8	5
Onion	146	48	0	1
Orange	303	95	2.0	30
Potato	1722	37	0.3	15
Strawberry	383	67	0.3	12
Tomato	359	23	0.3	26

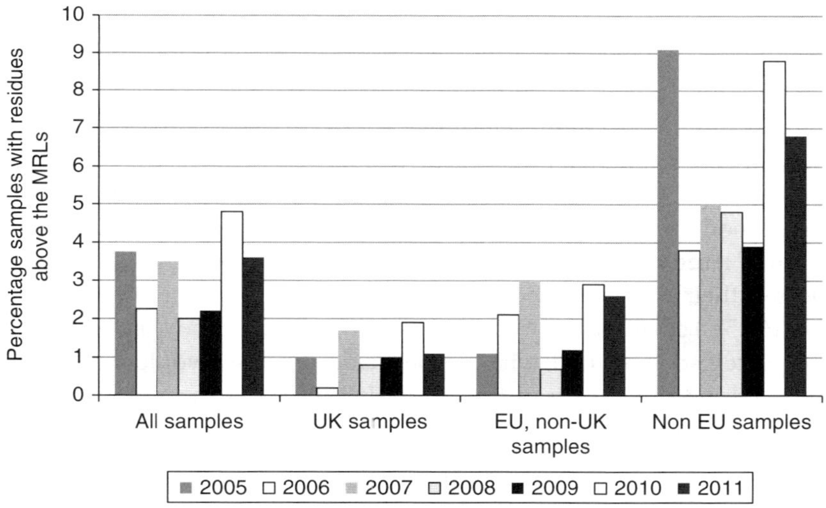

Fig. 7.6 Percentage of samples with residues above the MRL detected in the UK (Pesticides Forum Annual Report, 2012. Contains public sector information published by the Health and Safety Executive and licensed under the Open Government Licence. © Crown copyright).

insecticides constituted 99% of acceptable daily intake (ADI), with the insecticides – dimethoate and chlorpyrifos methyl, and fungicides – flusilazol, tebuconazole and captan exceeding the acute reference dose (ARfD). Overall it was considered that the occurrence of pesticide residues in apples could not be considered a serious public health problem. Similarly in studies on brassica crops the combined cumulative exposure to residues of chlorpyrifos were 0.777% of ADI, so not be considered to be a serious public health problem (Lozowicka et al., 2012), but an investigation into continuous monitoring and tighter regulation of pesticide residues is recommended.

Table 7.3 Acute risk exposure assessment from an EU coordinated programme (Nasreddine and Parent-Massin, 2002. Reproduced with permission of Elsevier)

Pesticide	Food	Maximum residue found (mg/kg)	ARfD (kg food/day)	97.5th percentile of consumption (mg/kg/day/kg bw)	Intake (mg/day/kg bw)	% ARfD
Chlorpyrifos	Oranges	0.55	0.1	0.235	0.00753	0.75
Endosulfan	Peaches	0.95	0.02	0.958	0.01062	53
Methidathion	Oranges	3.5*	0.01	0.235	0.00958†	96

* Determined on the whole unpeeled orange.
† Calculated with a variability factor of 7 and a factor of 0.1 for the edible portion.

Examples of acute risk assessment are shown in Table 7.3.

Although some studies have shown lower pesticide residues in organic produce, these crops are not always entirely free of pesticide contamination with residues sometimes found, albeit at levels lower than those found in conventional crops sprayed with a pesticide. This may be due to a previous history of contaminated soil or due to some spray drifting from a neighbouring farm. Benbrook and Baker (2014) have concluded that the organic standard should be a more precautionary risk-based approach rather than enforce a target aimed at no pesticide residue.

In the UK, Bradbury et al. (2014) examined the hypothesis that eating organic food might reduce the risk of breast cancer and other common cancers in a large prospective study of 623,080 middle-aged women. They reported their consumption of organic food as (i) never (30%); (ii) sometimes (62.7%) or (iii) usually or always eating organic food (just over 7%), as they perceived that food with no pesticide residues would be healthier. Over a period of 9.3 years, a total of 53,769 cases of cancer were reported, but they found little evidence for a decrease in the incidence of all cancers, except perhaps for non-Hodgkin lymphoma.

When 46 different fruits and vegetables imported from eight South American countries into Scandinavia in 2007 were examined for pesticide residues, 8.4% of samples contained pesticide residues above MRL, while out of a total of 724 samples, 19% of the samples no residues were found and the remaining 72% of samples contained pesticide residues at or below MRL. Thiabendazole, imazalil and chlorpyrifos were the pesticides most frequently found. Thirty-seven pesticides were found with frequencies higher that 1% in the samples, emphasising a need to monitor pesticide residues, especially in imported fruits and vegetables (Hjorth et al., 2011). This led to increased control for certain food products; data from Southeast Asia produce was reported by Skretteberg et al. (2015).

Jardim and Caldas (2012) have reported that less 3% of samples analysed in Brazil from 2001 to 2010 had residues above the MRL. Apple, papaya, sweet pepper and strawberry were the crops with the higher percentages of positive samples and almost half of the samples had residues of up to 10 different pesticides. The use of non-authorised active ingredients was also reported. The high residue on some fruit crops is in some cases due to late fungicide applications to allow longer storage. In contrast, prior to 2000, about 23% of samples of vegetable crops in India had residues of OP insecticides above their respective MRL values (Kumari et al., 2002), although pesticide use had been changing with more applications of pyrethroid insecticides. In China, 6% of samples in one survey had residues of two or more pesticides, with several OP insecticides detected above the MRL (Wang et al., 2013).

Organochlorine residues were detected using ion-trap GC–MS/MS spectrometry above the MRL in 5% of samples of strawberries from organic and IPM crops in Portugal (Fernandes et al., 2011) despite these insecticides not being used over the last 30 years due their persistence in the soil environment. Subsequent more extensive screening for 600 pesticides using other spectrometric analysis techniques revealed residues of other pesticides, including fluazifop-*p*-butyl (herbicide) and fenpropathrin (acaricide/insecticide) residues in IPM crops, but not in organic crops. However, all the residues were below the EU MRLs (Fernandes et al., 2014).

Up-to-date information relevant to the UK is available at www.pesticides.gov.uk by looking at the Expert Committee on Pesticide Residues in Food (PRiF) reports. Other information on MRLs and what pesticides may be used can be obtained by subscribing to the Food and Environment Research Agency (FERA) database 'Liaison', which is an on-line knowledge system that allows rapid identification of approved uses of pesticides in the UK. The proportion of samples with residues above the MRL on samples analysed in the UK is shown in Figure 7.6.

With the latest technique in analytical chemistry, it is now possible to detect and measure extremely small quantities of a pesticide within a large sample of the commodity, for example parts per billion. A residue below the level of detection is sometimes assumed to be just below the limit of quantification (LOQ), unless other evidence indicates a zero residue. Similar data are available from the web pages of some large food companies and supermarkets and other national authorities as well as the EU and USA, for example http://www.ams.usda.gov/science/pdp/Download.htm. Some residues may be bound in the food and cannot be extracted by standard residue analysis. While this has been considered to be of no toxicological concern, because this residue was not bioavailable, more recent studies have examined the possibility of correcting maximum residue levels of highly bioavailable-bound residues (Sandermann, 2004).

While agrochemical companies determine the MRL for with emphasis on the major crops, there are problems in setting values for minor crops. In the USA, an inter-regional research project (IR-4) was set up to provide data for minor crop farmers (Baron et al., 2003).

Before a pesticide can be registered the MRL would be carefully compared with the ADI, the ARfD and the expected intake of the food. If the

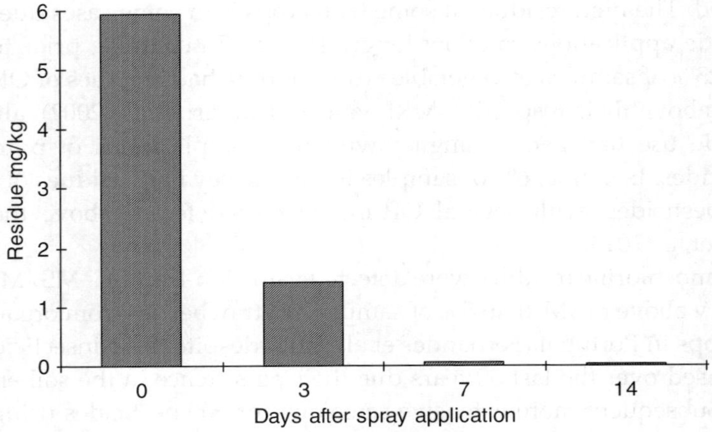

Fig. 7.7 Example of decline in residues following a pesticide application.

MRL value was too high, the pesticide would not be registered for use. The company who wished to market the pesticide may then do further trials with a lower dose or extended PHI, and providing it was still effective against the pest, it may be possible to register it with specific recommendations concerning the maximum dose permitted and when it may be applied. The MRLs in the UK are according to the Government regulations (Anon, 1999) as amended. If excess residues are found in a commodity, the producers can be fined and foreign imports banned.

An example of the decline in residue in a food crop is shown in Figure 7.7. In this example, the MRL might be set at 1.0 mg/kg. Some large food companies will also do their own residue analyses to ensure that their suppliers follow the codes of practice that they set (Table 7.4). Suppliers to supermarkets are encouraged to follow food assurance schemes, which are run as product certification schemes that are accredited by the United Kingdom Accreditation Service (UKAS). Examples are the Red Tractor and LEAF (Linking Environment and Farming) schemes. These schemes require that all those providing professional advice on pesticide usage on-farm must be a member of the BASIS Professional Register.

When a high proportion of tested samples contain residues, more detailed surveillance is carried out. One example in 2002 related to pears when 81% of UK samples and 66% of imported pears had pesticide residues of chlormequat, a gibberellin biosynthesis inhibitor thought to increase flowering and fruit yields, but growers no longer use it. Residues of fungicides, notably carbendazim, have also declined with changes in pesticide usage and storage practice. The increase in the amount of food imported into the UK from tropical countries has caused concern as there has been generally less regulation of pesticide use locally. One example of a problem is where high residues have been found on yams treated post-harvest with

Table 7.4 Residue analyses data from Marks and Spencer relating to tests from July to September 2008, 99 food samples were tested for an average of 175 different pesticide residues, equivalent to 17,325, individual tests

	Number of samples	Number of residues detected	Number with residues below legal limit	Number with residues above the legal limit
Vegetables	24	22	1	1
Fruits	22	4	17	1
Salads	29	17	12	
Organic	5	4	1	
Potatoes	6	2	4	
Other foods	13	11	2	
Total	99	60	37	2

a fungicide. Delcour et al. (2015) have developed a risk-based pesticide residue monitoring tool to prioritize the sampling of fresh produce.

MRLs for food commodities in international trade are set by the Codex Alimentarius Commission (Codex), established jointly by FAO and WHO as an international inter-governmental food standard organisation (Van Eck, 2004). Generally permitted legal limits for residues in food are based on the 'as low as reasonably achievable' (ALARA) principle. MRLs within the EU have now been harmonised, and extended to crops not necessarily grown within the EU, and this is anticipated to have an effect on the export crops of ACP (Africa, Caribbean and Pacific) regions. The database at http://ec.europa.eu/sanco_pesticides can be used to find out the MRL for different pesticides and crops. Where there is no appropriate data from GAP to derive an MRL, the limit of detection is used. This has resulted in some pesticides being withdrawn as manufacturers have decided that the cost of generating the necessary field and safety data are not justified by the market for their product. To assist these countries, there is a pesticide initiative programme (PIP) that provides protocols and advice to growers and exporters of fresh produce to assist them in complying with changing standards and regulations. By developing integrated pest management systems better timing of treatments and possibility of reduced dosage can reduce risk of a residue in the harvested produce (Fig. 7.8). In the UK fungicides are applied more on certain food crops than insecticides (Fig. 7.9). In an effort to reduce residues in strawberries, a research programme used a forecasting model to determine disease risk periods so that conventional fungicide use was restricted to pre-flowering and post-harvest use only. Sulphur or potassium bicarbonate used during flowering for disease control gave to acceptable levels of control while showing a significant reduction in the number of pesticide residues in the fruit. The integrated programme was particularly effective (Saville et al., 2013).

When the MRL data are published, newspapers often give headline prominence to the number of samples that exceed the MRL, even when it

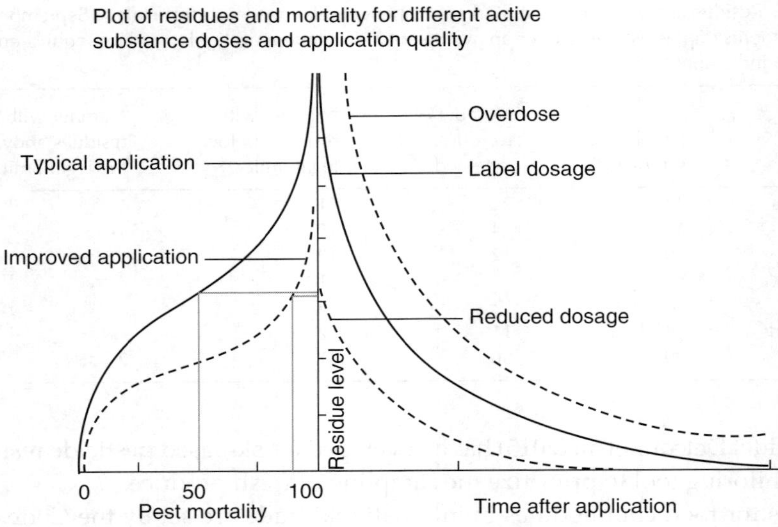

Fig. 7.8 Reduced dosage in field and subsequent decay of deposit to result in lower residue levels in harvested produce compared with label recommendation and over dosing.

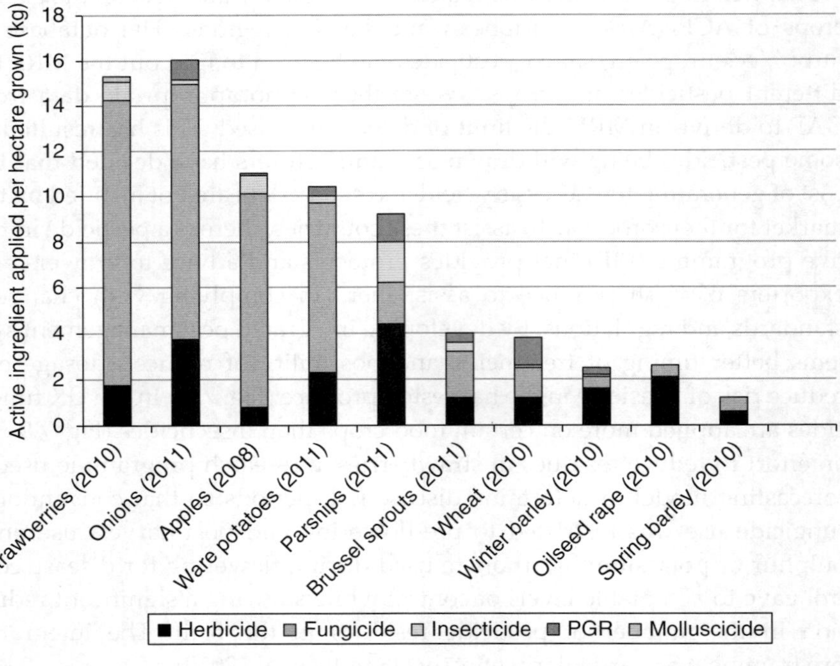

Fig. 7.9 Pesticides used on selected crops in UK on some of which there is major use of fungicides (Pesticides Forum Annual Report 2012. Contains public sector information published by the Health and Safety Executive and licensed under the Open Government Licence. © Crown copyright).

may be a small fraction of the samples analysed. Even when a sample slightly exceeds the MRL, there is seldom a health risk, as the MRL times the quantity of the food ingested will normally give a value far below the ADI. In most situations, it is also true that a person would not eat the same quantity of the same food daily for the whole of their life. To obtain some idea of the worst possible scenarios of residue intake, three estimates are used in risk assessments (Renwick, 2002). These are the theoretical maximum daily intake (TMDI), the national estimated daily intake (NEDI) and the national estimate of short-term intake (NESTI). Guidelines for predicting dietary intake of pesticide residues have been published by WHO (1997). These guidelines provide a method of reaching reasonable assurance that the intakes of pesticide residues for different populations do not exceed safety limits, and describe procedures that can be used by national authorities to predict the dietary intake of pesticide residues and decide the acceptability of MRLs from a public health point of view. Low et al. (2004) looked at published residue data and ranked results in various ways, concluding that there were no common trends because there is no single pesticide of particular concern from a consumer exposure viewpoint.

The use of DDT has been banned for many years, but a study in China showed that seafood from a coastal area was contaminated with DDT and the estimated daily intake (EDI) was higher in the coastal area than inland, although the value was below the EPA oral reference dose of 500 ng/kg bw/day (Man et al., 2013).

TMDI is not easy to determine because it is based on a high long-term consumption of a food and the MRL, corrected for loss of residue during transport, storage and processing or cooking prior to consumption. Where a food is processed or cooked, the intake must consider only the residue in the part of the food that is actually eaten. Where a residue is mainly on the outer surface of a food, washing or peeling it could reduce the residue intake. Washing with water, for example will remove some surface deposits, but a residue particularly of systemic pesticides can remain in the bulk of the food item. The amount that is removed from the surface will depend on the physico-chemical properties of the pesticide, but gentle rubbing of the surface by hand, while washing, will assist removal of deposits. Peeling or trimming vegetables and fruit can significantly remove surface residues (e.g. Kong et al., 2012; Han et al., 2015). Kaushik et al. (2009) review effects of different methods of food processing. A change in residue in a processed food compared to the raw agricultural commodity (RAC) is referred to as a transfer factor (Table 7.5). As an example, if olives contain a residue of 0.5 mg/kg and the extracted olive oil has 0.2 mg/kg, then the transfer factor is $0.2/0.5 = 0.4$. Transfer factors <1 indicate reduced residues, whereas >1 occurs if the residue is concentrated by the processing of the RAC. In determining transfer factors, samples containing residues similar to the MRL are needed to obtain measurable residues. However, registration of a product

Table 7.5 Some examples of transfer factors (Timme and Walz-Tylla, 2004. Reproduced with permission of John Wiley & Sons)

Crop	Pesticide	Process	Transfer factor
Apple	Captan	Ozone wash	0
		Juice	01–0.3
		Drypomace	2–4
	Pirimicarb	Washed	0.7
Banana	Tebuconazole	Peel	1–2
		Pulp	0.8–1
Buckwheat	Malathion	Noodles	0.4
Carrot	Chlorfenvinphos	Peeling/trimming	0.2
Orange	Profenophos	Peel	3
		Pulp	<0.1
Tomato	Pirimicarb	Washed	0.6
	Buprofezin	Juice	0.1
		Drypomace	34
Wheat	Bifenthrin	Bran	3–4
		Flour	0.3
		Bread	0.1
		Wholemeal flour	0.8–0.9
		Wholemeal bread	0.2–0.3

will be made only if the residue level is sufficiently low, that there is no need to wash or peel the food item.

Where the same pesticide may be used on different crops, the TMDI has to refer to the mean intake of the different foods (Table 7.6). TMDI is very much a theoretical value, but where it is less than the ADI, the possibility of the ADI being exceeded is extremely unlikely.

Changes in diet will influence the TMDI calculated for different regions of the world. In an example, given by WHO (1997), the TMDI for the herbicide 2,4-D varied from 7 to 50% of the ADI (Table 7.7).

A better estimate of the long-term intake of pesticide residues is derived from NEDI, which is the sum for all food commodities of the intake of the food commodity times the relevant residue level of the commodity and corrected for any change in residue level caused by processing or cooking. This residue level is not the MRL, but a median level is determined in supervised trials (STMR), when applying the maximum permitted dose under GAP. Where the median residue is below the level of detection (LOD), the LOD is used to calculate NEDI.

Where a crop is seasonal short-term intakes may be greater than the average. The risk is assessed using NESTI, derived from single day consumption data, but this is complicated by possible variation between samples of the commodity. A person may consume a higher residue on a single occasion or day, because certain food has a higher residue than average, or the person eats more food with residues in one day. Eating a large portion of food with a high residue would be the worst case. In practice a consumer is unlikely to eat more than one commodity, such as a carrot, which

Table 7.6 Example of TMDI calculation for an insecticide tebufenozide (WHO, 1997)

Commodity	MRL (mg/kg)	Diet (g/day)	TMDI
Grapes	0.5	18.02	0.0090
Husked rice	0.1	12.00	0.0012
Pome fruits	1	45.00	0.0450
Potato	0.5	0.00	0.12
Walnuts	0.05	1.00	0.0001
Total		0.18	

Percentage of ADI = 16%.

Table 7.7 Example of TMDI from different regions (WHO, 1997)

Region	Diet (g/day)*	TMDI	TMDI as percentage of ADI (%)
European	178.00	0.2744	46
Latin American	116.75	0.1974	33
Far Eastern	114.83	0.0961	16
Middle Eastern	327.25	0.1636	50
African	28.33	0.0447	7

*Based on barley, black cherries, citrus fruits, eggs #, maize #, meat #, milk products #, milks #, oats, potato, raspberries, rice #, rye, sorghum # and wheat. # MRL at LOD.

Table 7.8 Residue level mg/kg in the edible portion of a single unit of food commodities required to be eaten by a 60 kg person to ingest the ARfD

Commodity	Unit weight of edible portion	Acute reference dose		
		0.0008	0.003	0.1
Apple	127	0.4	1.4	47
Carrot	89	0.5	2.0	67
Peach	99	0.5	1.8	14
Potato	160	0.3	1.1	38
Tomato	123	0.4	1.5	49

happens to have a high residue on the same day. If the commodity is well mixed during processing, NESTI is calculated from the amount of the commodity eaten times the residue corrected for processing, for example peeling and then divided by the body weight. This value would then be compared with the appropriate ARfD (Table 7.8).

The amount and type of food consumed by individuals varies so estimates are made not only for adults, but also for infants, toddlers (1.5–4.5 years), children and adults. Certain foods eaten by adults have been excluded from the example in Table 7.9 that shows that the lower weight of younger persons is reflected by a higher NEDI.

As children have a higher rate of metabolism, less mature immune systems and eat different foods, consuming more food per body mass, than adults, their exposure to pesticides has caused concern. In the UK an extra safety factor was proposed in relation to infant foods (Schilter et al., 1996) to

Table 7.9 Changes in NEDI between adults and children

	Adult	Child	Toddler	Infant
Body weight (kg)	70.1	43.3	14.5	7.5
Mean NEDI* (mg/kg bw/day)	0.000047	0.000057	0.000091	0.000114
Total NEDI**	0.0006	0.0007	0.001	0.009

* These example values are the mean of several individual foods for one insecticide.
** Calculated NEDI based on amount eaten as shown by surveys of diet.

allow for the extra sensitivity of infants to toxicants, especially neurotoxins. In a study in the USA (Fenske et al., 2002), the diets of two small groups of children of pre-school age were sampled for 15 targeted organophosphate insecticides. In the 88 samples of food, that were divided into fruit and vegetables, beverages, processed foods and dairy products, 16 samples had a detectable residue of at least one OP, with two of these samples having residues of two insecticides. Only in one sample was the acute population-adjusted reference dose (aPAD) for chlorpyrifos (1.7 μg/kg/day) exceeded, whereas in all other cases the exposure was up to 0.24 μg/kg/day. Other routes of exposure were not assessed in this study, so the children could get a higher dose from contaminated surfaces in their house or garden.

Some of the residues found in crops may be due to treatment either shortly before harvest or during storage. Some crops are sprayed close to harvest with fungicides to reduce potential losses due to disease in marketing and storage, while insecticides may be applied in stores.

Problems have arisen due to variation within samples, although in making assessments it is usual to assume that the total pesticide residue measure in a bulk sample is derived from one unit of the bulked sample. Thus if a bulked sample of 20 carrots contained $\times$ mg/kg, it might all be in one carrot. When samples of fruit were examined, there was no correlation between the residue concentration or surface residue and the mass of apples (Ambrus, 2000). The view was that the residue distribution was most likely to be influenced by the size, shape and density of the plants and mode of application. Thus the variability in the initial spray deposit was a key factor influencing the ultimate residue in the harvested fruit. In contrast, following the discovery that residues in carrots could vary by up to 25 times that in a composite sample (Harris, 2000), experiments with carrots could not identify any single factor that was the cause of high residues in individual roots (Carter et al., 2000). The variations in daily intake of food that occur are not a normal part of consumer risk assessment, which is a limitation when there is an acute toxicity risk from a pesticide. However, probability distributions of residue in food consumed can be calculated, and using probabilistic modelling account is taken of the variability of detectable residues to give a realistic estimate of risks from short-term exposure (Hamey and Harris, 1999). Some of the residue problems which have occurred have been due to excessive doses being applied, poor calibration of equipment, too many sprays being applied or overdosing due to overlapping swaths. With integrated pest management the aim is to minimise the dose of insecticides to allow greater survival of natural enemies, and this also reduces the risk of residues remaining in the harvested produce (Fig. 7.8). Non-compliance of PHIs and spray drift from adjacent crops have also been implicated. These problems are potentially more serious in developing countries that lack the training in correct application of pesticides (Matthews et al., 2003). Efforts are being made to help developing countries, especially the Africa, Caribbean and Asian countries (ACP) through the PIP to meet the new standards for importing fresh produce into Europe. Retailers are using Assured Produce schemes that meet EUREGAP and similar standards.

Many of the residue problems of earlier decades have diminished as the persistent chemicals, such as the organo-chlorine insecticides have been banned and replaced by less persistent pesticides. In endeavouring to produce food at low cost and avoiding losses after harvest, some crops are treated close to harvest or following harvest. Such treatments tend to be the ones which result in a detectable residue in the produce, as the pesticide is not diluted by the growth of the plant, and has little time to be metabolised or degraded. There is therefore a conflict between the desire for chemical-free food and low cost (Foster et al., 2003).

Risk assessments of pesticides have been made in relation to individual chemicals, but a crop may be treated at different times with a range of pesticides. Where farmers do use different chemicals, for example an insecticide may applied with a fungicide at the same time to avoid spraying on two occasions or different herbicides are used in a mixture to treat a range of weed species, the pesticides have different modes of action and operate independently. As a harvested product may have the residue of more than one pesticide, 'The Committee on Toxicity' in the UK recently reviewed (Anon, 2002) the risks associated with mixtures of pesticides. It was concluded that there is evidence for limited exposure of humans to multiple residues and that such exposure occurs at low levels. An example of multiple residue is shown in Table 7.10, in which three of the chemicals, two insecticides and one acaricide exceeded their MRLs.

The main concern of the public was the possibility of adverse reactions due to the 'cocktail effect', but there is little evidence of the occurrence of such combined effects in humans. Nevertheless there is a need to assess the potential risks of combined exposures to multiple residues, particularly where the chemicals have the same mode of action (Boobis et al., 2008). They pointed out that there are a number of ways for cumulating toxicity, which are inter-related, and in increasing levels of complexity and refinement, are the hazard index, the reference point index, the Relative Potency Factor method and physiologically based toxicokinetic modelling.

Table 7.10 Example of residue of several pesticides found in one sample of Velcore beans, imported from Kenya (Data from Pesticide Residues Committee report on Sample 3238/2004, published in March 2005)

Pesticide	Type	Residue found (mg/kg)	MRL (mg/kg)
Cypermethrin	I	0.2	0.5
Dicofol	A	0.6*	0.02
Dimethoate	I	0.2*	0.02
Dithiocarbamate	F	0.2	1
Omethoate	I	0.1*	0.02
Profenofos	I	0.05	0.05
Propargite	A	0.1	CAC = 20
Tetradifon	A	0.1	No MRL

Data from Pesticide Residues Committee report on Sample 3238/2004 published March 2005.
A, acaricide; CAC, Codex Alimentarius Commission value used when there is no UK MRL; F, fungicide; I, insecticide.
* Exceeded MRL.

Where a few well-designed experiments have shown synergistic or antagonistic interactions or additive effects, they have occurred at high concentrations or exposure levels, which are probably unrepresentative of real-life exposures. In one study, groups of rats fed a diet containing different daily doses of chlorpyrifos also received daily doses of four other pesticides – alphacypermethrin, bromopropylate, carbendazim and mancozeb (Jacobsen et al., 2004). Co-administration of these pesticides did not enhance inhibition of acetylcholinesterase activity in plasma or the brain. Some effects were observed, where combinations had been administered, for example increased weight of the liver and thyroid, but further studies would be needed to ascertain which of the pesticides caused them.

Research is now focusing on the more vulnerable groups, such as young children, where the body's metabolic pathways are still developing to deal with the range of toxic compounds to which a body is exposed. Due to the lower body weight of children, any pesticide residue ingested would be more concentrated within the body.

Total diet studies have been mainly in the USA, Europe and Japan, but a few studies relate to tropical countries. Sawaya et al. (2000) reporting pesticide levels in Kuwait showed that in one cereal product fenitrothion did exceed the MRL sufficiently to warrant action. Most instances of pesticide poisoning due to eating food with high residues are rare, but when they do occur, they have been due to misuse of a pesticide on a crop, for which its use is not registered. Aldicarb on watermelons in California was one example of misuse. Deliberate use of a rodenticide on a food, intended as a rat control bait has also caused poisoning and mortality when the bait was eaten by humans. In the USA, a Community Duplicate Diet Methodology is proposed as a tool to provide more accurate assessments of intake of pesticides, which may enhance decisions for chemical regulation (Melnyk et al., 2014).

In France, Nougadère et al. (2012) reported on a total diet study in relation to chronic risk to pesticide residues. A further report describes a scoring method with different priority levels (Nougadère et al., 2011) to assess acute and chronic risks to the general population by monitoring dietary exposure to pesticide residues and identify food commodities and pesticides that need to be better monitored and/or regulated. In a recent study in which 522 pesticides and their metabolites were examined 87% scored the lowest priority level 1. Action was needed for the pesticides in level 6, namely carbendazim, dimethoate, dithiocarbamates and imazalil frequently found in fruits and vegetables to ensure consumer safety (Nougadère et al., 2014).

There is considerable concern among some people that food should have any residues of a pesticide in it. Should they be concerned? Even before any pesticides were developed food contained a range of other substances, which were not necessarily nutritious and some of these could be toxic. Plants have evolved an array of toxins as a defence against attacks by insects and pathogens and to deter grazing animals. The hotness of chilli peppers is due to capsaicin, which is anti-fungal agent and anti-feedant. A variety of potatoes, 'Lenape' planted by organic growers had to be withdrawn due to their high content of solanine and chaconine, toxic to man (Fenwick et al., 1990). There are monitoring programmes for some crops, such as potatoes to ensure that the safety level for natural toxins is not exceeded (Table 7.11). High yielding potato varieties have been selected to be resistant to certain pests, for example Maris Piper is resistant to some cyst nematodes, but still requires pesticide protection from aphids, slugs and late blight (Foster et al., 2003).

In the development of most of the major food plants, such as wheat, rice and maize, varieties have been selected with relatively few toxins so that they are palatable to us. Leaves of wild cabbage (*Brassica oleracea*), from which modern cabbage, broccoli and cauliflower have been bred, contain twice the amount of many glucosinolates as cultivated cabbage (Mithen et al., 1987).

Thus, many foods, that we do like, contain substances that we refer to a xenobiotics, that is they are foreign to our bodies and are not nutritious. The stimulant caffeine in coffee is a good example of this, and in the USA there is a legal limit of 6 mg of caffeine per liquid ounce in beverages (acute oral

Table 7.11 Examples of glycoalkaloid content in certain potato cultivars (Berry, 2004. Reproduced with permission of John Wiley & Sons)

Cultivar	Glycoalkaloid content (mg/100 g dry weight)
King Edward	80–120
Pentland Hawk	90–130
Epicure	110–140

Potatoes with >200 mg/100 kg dry weight may not be marketed.

LD_{50} 150 mg/kg of caffeine). However, as man has evolved, our bodies have co-evolved systems to break down small quantities of these xenobiotics so that they are harmless or are excreted. A person taking a painkiller such as paracetamol will benefit from taking the prescribed dose, but the body cannot cope adequately from an overdose and too many tablets kills the patient (Lappin, 2002). Even in the sixteenth century Paracelsus (1493–1541), a Swiss medical practitioner noted that the right dose differentiates a poison from a remedy.[1] So our body can cope with small amounts of pesticide in our diet and metabolise them. The problems occur with an overdose as in suicide attempts, or excessive exposure during application of pesticides. As described in earlier chapters, before a pesticide is registered and allowed to be sold, it is extensively tested to ensure that when applied as recommended, any residue in the crop at harvesting will not be hazardous to eat. In contrast some of the 'natural' foods would fail some of these tests (Ames et al., 1990).

Some of the unusual chemicals in some foods are considered to be therapeutic. The flavinoids are one group of plant polyphenols, some of which are thought to play a role in maintaining health, although others may be toxic. Among these antioxidant phytochemicals are the procyanidins, found in several foods such as apples, almonds, barley, grapes, tea, maize, cinnamon, cocoa, peanuts, wine and strawberries. These procyanidins may modulate key biological pathways in mammals. A high polyphenol diet has been shown in epidemiological studies to reduce the risk of coronary heart disease and stroke, by inhibiting oxidation of low-density lipoproteins (LDLs). It is claimed that the risk of atherosclerosis developing is by preventing LDL (regarded as bad cholesterol) building up plaque in the arteries, while increasing the good cholesterol. Cocoa, chocolate, green tea, grapes, apples and red wine contain flavonoids, which are a specific subclass of these compounds, which have also received attention as being beneficial to health. According to Johansson et al. (2014) the content of nutritional beneficial compounds in foods was not influenced by organic cultivation. Culliney et al. (1993) give a more detailed account of the question of natural toxicants in food.

According to the UK Food Standards Agency, fruit and vegetables should make up about a third of the food eaten each day. It is also important to eat a variety of fruits and vegetables with 'five-a-day' as a good, achievable target. One portion is considered to be 80 g. FSA consider that the risk to health from eliminating fruit and vegetables from the diet would far outweigh the risks posed by possible exposure to pesticide residues. As an example if a 55 kg weight person ate 80 g of cauliflower curds, the theoretical maximum residue contribution (TMRC) of an insecticide (indoxacarb) calculated from residues observed on day 0 samples would be 20.8 and 36.8 μg for the recommended and double the recommended dosages. These values compare with the acceptable daily intake (ADI) 0f 550 μg indicating a very safe margin (Takkar et al., 2011), especially as the residues dissipated to below the level of quantification in 7 days.

Farmers generally only use pesticides when it is economically justified to protect their crops and thus achieve a higher yield of marketable produce. Market forces with 'assured produce' schemes and similar programmes operated by reputable supermarkets and wholesalers ensure that if a farmer is to remain in business, every effort will be made to ensure that any residue will be below the MRL. An extensive return to 'organic' produce without carefully regulated pesticides will inevitably raise the costs of commodities and increase the presence of lower quality produce with a short shelf-life.

The public is concerned about the presence of pesticide residues in foods. This chapter has shown that by sampling and very sophisticated analytical techniques, it is now possible to detect chemicals at extremely low quantities. The mere presence of these is not any indication of a risk to health unless they far exceeded the MRLs, a level that confirms that the pesticide has been applied according to good agricultural practice. Regular checks ensure that the residues found in our foods are well below the level that would cause concern.

Note

1 What is there that is not poison? All things are poison and nothing is without poison. Solely, the dose determines that a thing is not a poison.

References

Ambrus, A. (2000) Within and between field variability of residue data and sampling implications. *Food Additives and Contaminants* **17**, 519–537.

Ames, B.N., Profet, M. and Gould, L.W. (1990) Nature's chemicals and synthetic chemicals: comparative toxicology. *Proceedings of the National Academy of Science of the USA* **87**, 7782–7786.

Anon (1999) *The Pesticides (Maximum Residue Levels in Crops, Food and Feeding Stuffs) (England and Wales) Regulations 1999*, Statutory Instruments. Available at: http://www.legislation.gov.uk/uksi/1999/3483/contents/made (accessed on 8 May 2015).

Anon (2002) Risk Assessment of Mixtures of Pesticides and Similar Substances. Committee on Toxicity, The Food Standards Agency, London.

Baron, J.J., Kukel, D.L., Holm, R.E., Hunter, C., Archambault, S. and Boddis, W. (2003) Co-operative facilitation of registrations of crop protection chemicals in fruits, vegetables and other speciality crops in the United States and Canada. *The BCPC International Congress – Crop Science and Technology* **2003**, 583–588.

Benbrook, C.M. and Baker, B.P. (2014) Perspective on dietary risk assessment of pesticide residues in organic food. *Sustainability* **6**, 3552–3570.

Berry, C. (2004) Explaining the risks. In: Hamilton, D. and Crossley, S. (eds.) *Pesticide Residues in Food and Drinking Water: Human Exposure and Risks*. Wiley, Chichester, pp. 339–351.

Boobis, A.R. Ossendorpb, B.C., Banasiak, U., Hamey, P.Y., Sebestyen, I. and Morettof, A. (2008) Cumulative risk assessment of pesticide residues in food. *Toxicology Letters* **180**, 137–150.

Bradbury, K.E., Balkwill, A., Spencer, E.A., Roddam, A.W., Reeves, G.K., Green, J., Key, T.J., Beral, V., Pirie, K. and the Million Women Study Collaborators. (2014) Organic food consumption and the incidence of cancer in a large prospective study of women in the United Kingdom. *British Journal of Cancer*. **110**, 2321–2326.

Carter, A.D., Fogg, P. and Beard, G.R. (2000) Investigations into the causes of residue variability on carrots in the UK. *Food Additives and Contaminants* **17**, 503–509.

Chatzicharisis, I., Thomidis, T., Tsipouridis, C., Mourkidou-Papadopoulou, E. and Vryzas, Z. (2012) Residues of six pesticides in fresh peach—nectarine fruits after preharvest treatment. *Phytoparasitica* **40**, 311–317.

Culliney, T.W., Pimentel, D. and Pimental, M.H. (1993) Pesticides and natural toxicants in foods. In: Pimental, D. and Lehman, H. (eds.) *The Pesticide Question – Environment, Economics and Ethics*. Chapman and Hall, New York, pp. 1126–1150.

Delcour, I., Rademaker, M., Jacxsens, L., De Win, J., De Baets, B. and Spanoghe, P. (2015) A risk-based pesticide residue monitoring tool to prioritize the sampling of fresh produce. *Food Control* **50**, 690–698.

EFSA (2013) The 2010 European Union Report on pesticide residues in food. *EFSA Journal*, **11**, 3130.

European Commission (1998) Monitoring of Pesticide Residues in Products of Plant Origin in the European Union Report 1996. European Commission, Brussels, p. 15.

Fenske, R.A., Kedan, G., Lu, C., Fisker-Andersen, J.A. and Curl, C.L. (2002) Assessment of organophosphorous pesticide exposures in the diets of preschool children in Washington State. *Journal of Exposure Analysis and Environmental Epidemiology* **12**, 21–28.

Fenwick, G.R., Johnson, I.T. and Hedley, C.I. (1990) Toxicity of disease-resistant plant strains. *Trends in Food Science and Technology* **1**, 23–25.

Fernandes, V.C., Domingues, V.F., Mateus, N. and Delerue-Matos, C. (2011) Organochlorine pesticide residues in strawberries from integrated pest management and organic farming. *Journal of Agricultural and Food Chemistry* **59**, 7582–7591.

Fernandes, V.C., Lehotay, S.J., Geis-Asteggiante, L., Kwon, H., Mol, H.G.J., van der Kamp, H., Mateus, N., Domingues, V.F. and Delerue-Matos, C. (2014) Analysis of pesticide residues in strawberries and soils by GC–MS/MS, LC–MS/MS and two-dimensional GC–time-of-flight MS comparing organic and integrated pest management farming. *Food Additives and Contaminants* **31**, 262–270.

Foster, G.N., Atkinson, D. and Burnett, F.J. (2003) Pesticide residues – better early than never? The BCPC International Congress – Crop. *Science and Technology* **2003**, 711–718.

Hamey, P.Y. and Harris, C.A. (1999) The variation of pesticide residues in fruits and vegetables and the associated assessment of risk. *Regulatory Toxicology and Pharmacology* **30**, S34–S41.

Han, J-L., Fang, P., Xu, X-M., Zheng, X.L., Shen, H.T. and Ren, Y-P. (2015) Study of the pesticides distribution in peel, pulp and paper bag and the safety of pear bagging. *Food Control* **54**, 338–346.

Harris, C.A. (2000) How the variability issue was uncovered: history of the UK residue variability findings. *Food Additives and Contaminants* **17**, 401–495.

Hjorth, K., Johansen, K., Holen, B., Andersson, A., Christensen, H.B., Siivinen, K. and Toome, M. (2011) Pesticide residues in fruits and vegetables from South America e A Nordic project. *Food Control* **22**, 1701–1706.

Hyder, K. and Travis, K.Z. (2003) Maximum residue levels: a critical investigation. The BCPC International Congress – Crop. *Science and Technology* **2003**, 569–574.

Jacobsen, H., Ostergaard, G., Lam, H.R., Poulsen, M.E., Frandsen, H., Ladefoged, O. and Meyer, O. (2004) Repeated dose 28-day oral toxicity study in Wistar rats with a mixture of five pesticides often found as residues in food: alphacypermethrin, bromopropylate, carbendazim, chlorpyrifos and mancozeb. *Food and Chemical Toxicology* **42**, 1269–1277.

Jardim, A.N.O. and Caldas, E.D. (2012) Brazilian monitoring programs for pesticide residues in food – results from 2001 to 2010. *Food Control* **25**, 607–616.

Johansson, E., Hussain, A., Kuktaite, R., Andersson, S.C. and Olsson, M.E. (2014) Contribution of organically grown crops to human health. *International Journal of Environmental Research and Public Health* **11**, 3870–3893.

Kaushik, G., Satya, S. and Naik, S.N. (2009) Food processing a tool to pesticide residue dissipation – a review. *Food Research International* **42**, 26–40.

Kong, Z., Shan, W., Dong, F., Liu, X., Xu, J., Li, M. and Zheng, Y. (2012) Effect of home processing on the distribution and reduction of pesticide residues in apples. *Food Additives & Contaminants: Part A*, **29**(8), 1280–1287.

Kumari, B., Madan, V.K., Kumar, R. and Kathpal, T.S. (2002) Monitoring of seasonal vegetables for pesticide residues. *Environmental Monitoring and Assessment* **74**, 263–270.

Lappin, G. (2002) Chemical toxins and body defences. *Biologist* **49**, 33–37.

Low, F., Lin, H-M., Gerrard, J.A. Cressey, P.J. and Shaw, I.C. (2004) Ranking the risk of pesticide dietary intake *Pest Management Science* **60**, 842–848.

Lozowicka, B. (2015) Health risk for children and adults consuming apples with pesticide residue. *Science of the Total Environment* **502**, 184–198.

Lozowicka, B. Jankowska, M. and Kaczynski, P. (2012) Pesticide residues in Brassica vegetables and exposure assessment of consumers. *Food Control* **25**, 561–575.

Man, Y.B., Chan, J.k.Y., Wu, S.C., Wong, C.K.C. and Wong, M.H. (2013) Dietary exposure to DDTs in two coastal cities and an inland city in China. *Science of the Total Environment* 463–464, 264–273.

Matthews, G.A., Dobson, H.M. Wiles, T.L. and Warburton, H. (2003) The Impact of Pesticide Application Equipment and its Use in Developing Countries, with Particular Reference to Residues in Food, Environmental Effects and Human Safety. FAO, Rome.

Melnyk, L.J., Xue, J., Brown, G.G., McCombs, M., Nishioka, M. and Michael, L.C. (2014) Dietary intakes of pesticides based on community duplicate diet samples. *Science of the Total Environment* 468–469, 785–790.

Mithen, R.F., Lewis, B.G., Heany R.K. and Fenwick, G.R. (1987) Glucosinolates of wild and cultivated *Brassica* species. *Phytochemistry* **26**, 1969–1973.

Nasreddine, L. and Parent-Massin, D. (2002) Food contamination by metals and pesticides in the European Union. Should we worry? *Toxicology Letters* **127**, 29–41.

Nougadère, A., Reninger, J-C., Volatier, J-L. and Leblanc, J-C. (2011) Chronic dietary risk characterization for pesticide residues: a ranking and scoring method integrating agricultural uses and food contamination data. *Food and Chemical Toxicology* **49**, 1484–1510.

Nougadère, A., Sirot, V., Kadar, A.,Fastier, A., Truchot, E., Vergnet, C., Hommet, F., Baylé, J., Gros, P. and Leblanc, J.C. (2012) Total diet study on pesticide residues in France: levels in food as consumed and chronic dietary risk to consumers. *Environment International* **45**, 135–150.

Nougadère, A., Merlo, M., Héraud, F., Réty, J., Truchot, E., Vial, G., Cravedi and Leblanc, J-C. (2014) How dietary risk assessment can guide risk management and food monitoring programmes: the approach and results of the French Observatory on Pesticide Residues (ANSES/ORP). *Food Control* **41**, 32–48.

Renwick, A.G. (2002) Pesticide residue analysis and its relationship to hazard characterisation (ADI/ARfD) and intake estimations (NEDI/NESTI) *Pest Management Science* **58**, 1073–1082.

Sandermann, H. (2004) Bound and unextractable pesticide plant residues: chemical characterization and consumer exposure. *Pest Management Science* **60**, 613–623.

Saville, R., Berrie, A., Fitzgerald, J., Jay, C., Roberts, H., Wedgwood, E., Xu, X. and Cross, J. (2013) Minimising pesticide residues in strawberry through integrated pest, disease and environmental crop management. *IOBC-WPRS Bulletin* **91**, 431–438.

Sawaya, W.N., Al-Awadhi, F.A., Saeed, T., Al-Omair, A., Husain, A., Ahmad, N., Al-Omirah, H., Al-Zenki, S., Khalafawi, S., Al-Otaibi, J and Al-Amiri, H. (2000) Dietary intake of organophosphate pesticides in Kuwait. *Food Chemistry* **69**, 331–338.

Schilter, B., Renwick, A.G. and Huggett, A.C. (1996) Limits for pesticide residues in infant foods: a safety-based proposal. *Regulatory Toxicology and Pharmacology* **24**, 126–140.

Skretteberg, L.G., Lyrån, B., Holen, B., Jansson, A., Fohgelberg, P., Siivinen, K., Andersen, J.H. and Jensen, B.H. (2015) Pesticide residues in food of plant origin from Southeast Asia – A Nordic project. *Food Control* **51**, 225–235.

Takkar, G. Sahoo, S.K., Singh, G., Mandal, K., Battu, R.S. and Singh, B. (2011) Persistence of indoxacarb on cauliflower (*Brassica oleracea* var. *botrytis*. L.) and its risk assessment. *American Journal of Analytical Chemistry* **2**, 69–76.

Timme, G. and Walz-Tylla, B. (2004) Effects of food preparation and processing on pesticide residues in commodities of plant origin. In Hamilton, D. and Crossley, S. (eds.) *Pesticide Residues in Food and Drinking Water: Human Exposure and Risks*. Wiley, Chichester, pp. 121–148.

Van Eck, W.H. (2004) International standards: the international harmonization of pesticide residue standards for food and drinking water. In Hamilton, D. and Crossley, S. (eds.) *Pesticide Residues in Food and Drinking Water: Human Exposure and Risks*. Wiley, Chichester, pp. 295–338.

Wang, S., Wang, Z., Zhang, Y., Wang, J. and Guo, R. (2013) Pesticide residues in market foods in Shanxi Province of China in 2010. *Food Chemistry* **138**, 2016–2025.

WHO (1997) Guidelines for Predicting Dietary Intake of Pesticide Residues. WHO/FSF/FOS/07.7. WHO, Geneva.

8 The future of pesticides

As the global population is estimated to increase to 9 billion by 2050, production of food will have to increase in a situation where more people are migrating to cities and changes in climate predict more extreme flooding and droughts. Even without an increasing population, too many people remain hungry or suffer from micronutrient deficiencies, so higher food production is essential with higher crop yields. Opening up more agricultural land without destruction of forests and other natural habitats of wildlife is not an option. Higher productivity can only be achieved by developing improved crop varieties and new techniques of crop protection. Unfortunately many people and governments have rejected growing genetically modified crops, which offer some hope of increasing production with drought resistant and more nutritious cultivars. Increasingly the GM crops have several different genes stacked in one variety. Competition between commercial companies with the new biotechnology is introducing genes expressing different innovative traits, such as herbicide tolerance as well as protection from insect pests. Combination of two *Bacillus thuringiensis* (Bt) proteins – Cry1F and Cry1Ac – in cotton plants extended control of lepidopteran cotton pests, thus providing season-long protection. Growing these new cultivars and applying pesticides, integrated with other changes in farming practices are just some of the ways in which farmers will be able to combat pests and diseases and achieve the higher yields needed (Fig. 8.1). GM crops incorporating a pesticide are sometimes referred to as 'plant-incorporated protectants' (PIPs).

Resistance to growing GM crops has been particularly vocal in Europe. In the UK, large-scale trials were carried out to assess the effect of GM crops on biodiversity. Large plots of genetically modified herbicide-tolerant (GMHT) crops of maize, spring-sown oilseed rape and sugar beet were grown on large plots alongside equivalent commercial crops, treated as recommended with herbicides. The GMHT crops were sprayed with glyphosate (beet) or glufosinate-ammonium (maize and oilseed rape), when weeds were present and likely to reduce yield. Detailed records of not only the plants/weeds were kept but also observations were made of insects, including butterflies, and other wildlife in both plots at each site and in field margins (Champion et al., 2003; Squire et al., 2003). In these

Pesticides: Health, Safety and the Environment, Second Edition. G.A. Matthews.

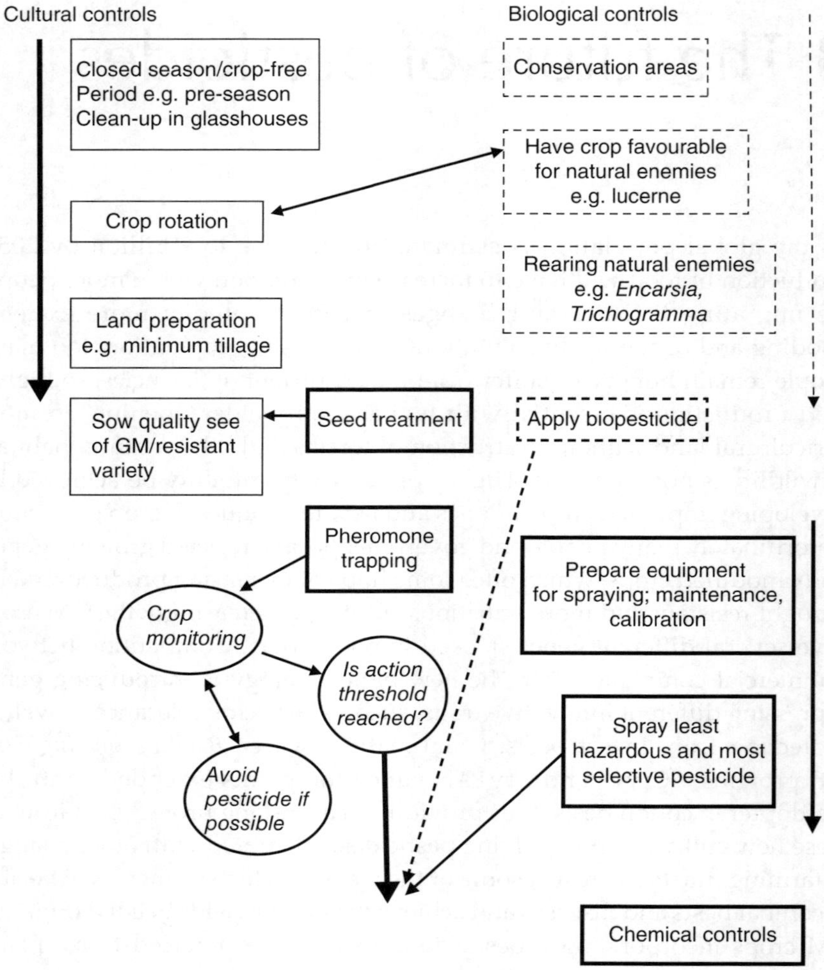

Fig. 8.1 Outline of different control tactics in an IPM programme.

three crops herbicides did not have any major impact on weed diversity, except briefly immediately after a treatment. Weed seed densities were lower in GMHT beet and rape crops, whereas despite more weeds in GMHT maize, seed returns were low irrespective of treatment. These effects in GMHT crops compounded over several seasons would, it was considered, decrease populations of some arable weeds significantly in beet and rape fields, although the reverse may occur in maize (Heard et al., 2003). The changes in weeds could have an impact on many invertebrates, small mammals and birds, whose populations interact with availability of their food supplies. Weed seed feeding carabid populations were smaller in GMHT beet and rape, but larger in GMHT maize, while collembola increased with GMHT crop management (Brooks et al., 2003).

In the row crops, no attempt was made to modify the application of pesticides. Instead of a band treatment leaving some weeds in the middle of the inter-row to be controlled later by tillage, the whole area was sprayed with a broad-spectrum herbicide. Leaving a strip of weeds on the inter-row until a later date was not studied although in a separate trial different dates of applying herbicide was investigated in GMHT sugar beet. Leaving weeds too long in the intra-row did compete with the sugar beet and reduced yields. In Denmark,the weed flora and arthropod fauna was denser and more diverse in GMHT fodder beets in early and mid-summer than conventional beets (Strandberg et al., 2005). Since these trials, the view of UK government has become more acceptable to GM crops, with a Minister arguing that GM crops have significant benefits for farmers, consumers and the environment, with future developments offering wonderful opportunities to improve human health. Countries growing Bt crops report positive health and environmental benefits by adoption of the new technology (Abedullah and Qaim, 2015; Huang et al., 2015). In contrast some countries continue to ban growing GM crops, following the views of organisations, such as Friends of the Earth and Greenpeace, while the global area of commercial GM crops, now grown in 28 countries, had expanded to 170.3 million ha by 2012 (James, 2012).

Weed management

Despite the development of hoeing equipment that can be controlled more accurately and by using in-line vision and GPS systems, the movement of the soil does not always suppress the weeds and can disturb roots of the crop and increase the danger of erosion and soil degradation. Furthermore the costs of weeding, either by manual or mechanical methods has risen, so increasingly farmers have adopted the application of herbicides to minimise the competition for nutrients in the soil. About half of all commercial pesticides are now herbicides, as they provide a less-expensive way of suppressing weeds.

Development of herbicide-tolerant GM crops allows a broad spectrum herbicide to control weeds later, after the crop is established, but before competition with the crop causes yield loss. Unfortunately the initial development of herbicide-tolerant crops has relied too much on one herbicide, glyphosate, so it has been overused with the consequence of large areas of weeds which have been selected with resistance to it. Green and Owen (2011) discuss the need to diversify weed management tactics and discover herbicides with new modes of action, so that different chemicals can be rotated to reduce selection of resistant weed populations. This should also be combined with other crops so that a range of herbicides can be used. New varieties of soyabean and maize have now been developed that are tolerant of 2,4-D herbicide and dicamba as well as glyphosate to extend their activity and control weeds resistant to glyphosate. The commercialization of the 2,4-D-tolerant crops has been resisted due to the perception that farmers will be applying 'Agent Orange', a herbicide that was used extensively during the Vietnam war. The toxicity of Agent Orange was due to the inclusion of a similar herbicide 2,4,5-T with dioxins and not 2,4-D. However other concerns are due to the volatility of 2,4-D which can result in damage of susceptible crops downwind due to spray and vapour drift, but the manufacturers claim this is no longer a factor due to a new formulation of the herbicide, using 2,4-D choline, which is a quaternary ammonium salt instead of the 2,4-D amine or ester formulations.

As ploughing of fields increases the possibility of soil erosion, especially on sloping lands with periods of heavy rainfall, there is now greater acceptance of zero (no-till) or minimum tillage systems. These tillage systems do not rely on the weed management function of the plough and also can only make reduced use of mechanical weeding like hoeing or cultivating, so herbicide use is a fundamental part of the system. Lack of ploughing has the advantage of preserving the environment of many soil organisms such as earthworms. No-till systems were initially depended on the application of paraquat which has a rapid contact action on foliage and the spray reaching the soil is adsorbed reducing the risk of movement from the treatment area. Concerns about its misuse as it is highly toxic if ingested, have led to its replacement by glyphosate, which is slow acting. It is very mobile in water, but in the soil environment absorbed and resistant to degradation. However 'free', glyphosate is quickly degraded by biological processes. In advanced no-till systems using permanent mulch covers of the soil, herbicides do not even reach the soil contrary to conventional farming, where a considerable part of the herbicides end up on bare soil surfaces. As soils under no-till produce less run-off, less leaching and are higher in organic matter, glyphosate does normally not create an environmental problem (Schuette, 1998; Jansen, 1999; Ruiz et al., 2001). As for any pesticide, herbicides have to be rotated with other products and other, non-chemical methods of weed management, in order to avoid herbicide resistance or the accumulation of the products in the environment. This applies even in plantation crops and no-tillage systems, where alternative weed control methods would include cover crops, mulch cover, mechanical controls like slashing or rolling (Neto, 1993). Chemical weed control is an important complement in no-tillage systems and has contributed to their increased popularity.

Sowing of GM oil-seed rape (Canola) expressing resistance to certain herbicides, especially glyphosate, has increased to over 75% of the world-wide area. This has decreased herbicide use by 40% as the new herbicide-resistant varieties require only one or two applications of the single broad-spectrum herbicide, whereas more traditional unmodified crops need several applications of herbicides, often applied as mixtures. Application of herbicides to GM herbicide resistant crops is after crop establishment so they can be applied more accurately to specific weed-infested areas, thus providing savings for the farmer and net benefits to the environment. Post crop establishment application of herbicides will be favoured by many small-scale farmers who have avoided investment in pre-emergence herbicides where rainfall is often erratic at the start of the season.

In the tropics where sufficient labour has been considered available for hand weeding, the trend will inevitably be towards more herbicide use as people migrate to towns or, where AIDS has had an impact on people and they are too ill to do hard work.

Disease management

Much has been done and will continue to be needed in terms of breeding disease-resistant cultivars. A change to a resistant cultivar can be relatively easy with annual crops, changes in perennial crops inevitably has to be over a longer period. Even with some level of resistance, many major crops suffer from infections of pathogens so require protection with fungicides. The production of high-quality produce that can be effectively stored is one situation where use of a fungicide at some stage prior to or at harvesting is often needed. More is now known about conditions that favour some of the important pathogens, so that by careful monitoring of temperature, rainfall, humidity and leaf wetness, the timing of a fungicide application, if needed, can be more accurate. Unfortunately, resistance to many fungicides has already occurred so resistance management strategies are essential.

Insect management

One clear advantage of incorporating a gene into plants that increases its resistance to insect pests is that there should be less spraying of insecticides or even no need to apply certain pesticides to the crop. In one sense the use of gene expressing a toxin within the plants such as the BT gene is a novel application technique, while also being a form of varietal resistance. Early data were not entirely clear as the expression of the Bt toxin allowed some targeted insect pests to survive, but as the knowledge in gene transfer has improved, so far the GM Bt crops have enabled farmers to reduce their use of insecticides against lepidopterous pests, notably the bollworms and especially *Helicoverpa* spp., although some sprays may be needed for 'sucking' pests such as aphids. Where fewer broad spectrum insecticides are applied against lepidopterans (e.g. bollworms), there is greater survival of predators,

(a)

(b)

Fig. 8.2 Monitoring insect pest populations. (a) Scouting cotton (Photograph by Graham Matthews), (b) trapping fruit flies (Photograph by Graham Matthews), (c) pheromone trap (Exocect) and (d) inside pheromone trap (Exocect).

so biological control has a better chance against the sucking pests. Crop monitoring, which is simple, cost effective and relatively quick to do, is needed in IPM to ensure other pests do not reach an economic level that justifies a spray application (Fig. 8.2).

(c)

(d)

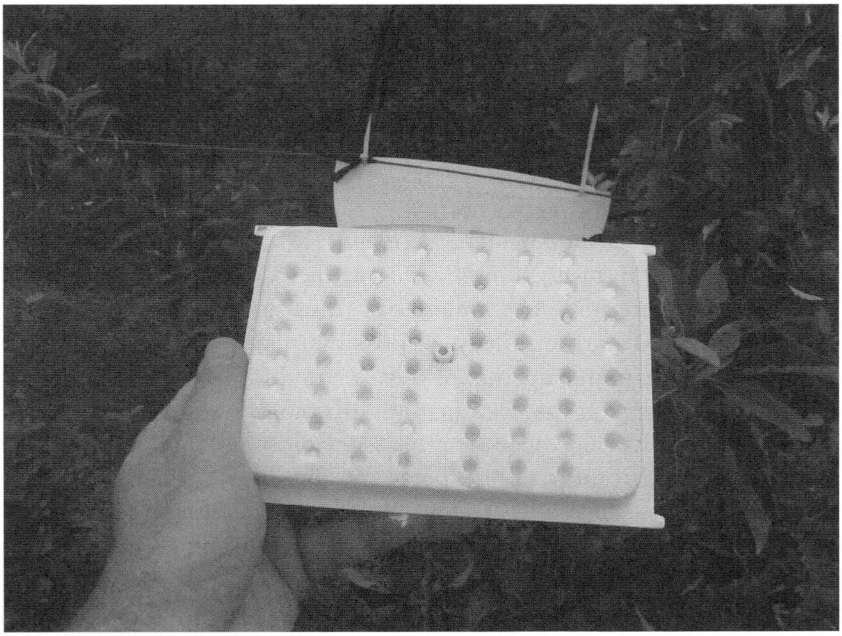

Fig. 8.2 *(Continued)*

This reduction in 'bollworm' sprays on cotton has already had an impact. In China, farmers were spraying sometimes on over 18 occasions per season, using small poor-quality knapsack sprayers, and often using insecticides classified by WHO as the most toxic. Hossain et al. (2004) referred to the period 1992–1996 when there were on average 54,000 cases of poisoning of farmers or their workers each year, 490 cases being fatal, but since the introduction of Bt cotton only 9% of those farmers growing it have reported poisoning, in contrast to 33% of farmers who grow non-Bt cotton. The view was that rather than criticising the adoption of GM crops due to perceptions on some speculative harmful effects, policy-makers should note the positive benefits of reducing the known risks of poisoning when using highly toxic pesticides.

Already since 1996, over 75% of the cotton area in the USA is now sown with GM varieties. In many other countries, including China and more recently India have recognised the value of GM cotton and increased the area with these new varieties. In Africa, GM cotton was initially tried in South Africa and from 2008 also in Burkina Faso.

GM maize has been grown with Bt toxins to control the corn rootworms which destroy corn roots, thereby reducing yield, and also cause plants to fall over during strong wind and rain, due to loss of roots, making harvest difficult. Unfortunately growing Bt maize to control corn root worm has already led to signs that the pest has become resistant to two kinds of Bt toxins included in corn – mCry3Bb1 and mCry3A.

IPM

As noted earlier integrated pest management (IPM) is now an integral part of EU policy on crop protection. IPM has attracted many definitions since it was originally conceived as a means of utilising different control techniques together as harmoniously as possible, using pesticides as a last resort (Fig. 8.1). To many, the aim of IPM is to avoid pesticides completely and essentially grow crops organically with biological and cultural control of pests. IPM has also been largely associated with endeavours to control either a specific pest or the pests on one crop, and often with an entomological bias. In reality, there is a need for cooperation within an agro-ecological area to adopt pest management covering all the crops and their major pests, in the widest context, within the whole area. Many key polyphagous pests attack many different crops and non-crop host plants within an area, yet historically there has been little attempt to combine efforts of different farms to work together against the pests. As an example, whiteflies, *Bemisia tabaci* will infest a wide range of plants affecting horticultural crops such as tomatoes and other field crops such as cotton in a country yet the efforts in the horticultural and agricultural industries have not been integrated.

IPM has to be understood in a wider context of the entire cropping system and environment, to avoid the use of pest management practices that, although being non chemical, have still a strong negative impact on the environment, such as burning of crop residues or ploughing.

Farmers find that outbreaks of pests can occur so intensively and sometimes over extensive areas, that reliance on non-pesticide controls fails. Much depends on the crop, the cultivar and where it is being grown, as climatic conditions can keep pest populations low in the areas with severe winters, although fewer pesticides are now available in the EU due legislation. Even with genetically modified crops, some pesticide use may still be needed, for example as a seed treatment. In practice pesticides will remain an important tool for farmers to maintain high yields. Their use, however, will become more selective both temporally and spatially, as farmers recognise the need to avoid blanket spraying of chemicals on a calendar schedule. Some pest may be controlled by using insecticides in conjunction with an attractant (food or pheromone). Routine walking the crop and scouting for pests has been advocated in many situations, so that a spray is only applied when strictly needed to avoid pest populations exceeding an economic threshold level. Patch spraying according to maps of weeds has also provided a means of limiting the use of herbicides to parts of fields where specific weeds occur.

As pointed out by the agrochemical industry, in addition to making agriculture more efficient and productive on a limited area of land, research by industry is helping to conserve and enhance biodiversity. This is by promoting systems such as IPM or rather Integrated Crop Management (ICM) throughout the world (sometimes referred to as IP^2M – Integrated Pest and Production Management). In broadening the scope beyond the management of pests, ICM encourages protection of natural wildlife habitats within and around the farm. Globally the aim is establish a network of protected areas around the world as stipulated by the Convention on Biological Diversity. These areas include temporary (e.g. uncut field margins) and permanent conservation areas, within farming areas.

In some cases, some pest can be controlled by releasing parasitoids or predators (Fig. 8.3), while strip management or under sowing (Fig. 8.4) can conserve and encourage natural enemies, and sometimes reduce pest population growth and obviate the need for an insecticide spray. Mensah (1999) gives examples of habitat diversity with strips of lucerne alongside cotton, while Levie et al. (2005) gives an example of aphid control. Many of these options have already been taken up by various schemes such as Linking Environment and Farming (LEAF) and some Farm Assurance Schemes which have evolved by close partnership between farming and those marketing farm produce (Fig. 8.5).

(a)

(b)

Fig. 8.3 (a and b) Releasing a biological control agent (Syngenta).

Traditional plant breeding

Plants contain natural pesticides to protect them from most insects and pathogens that attack them. To provide food suitable for humans, farmers for centuries have retained seeds from good plants, which had survived pests and diseases, for sowing the following season, but

Fig. 8.4 Organic crop inter-sown with clover (Photograph by Graham Matthews).

still have the qualities that make them suitable to eat. Plant breeding began in the nineteenth century when scientists made specific crosses between plants and selected those with the most useful traits to improve pest and disease resistance to achieve higher yields of better quality produce. It has been shown that even with partial resistance to a pest or pathogen, crops may require less protection with pesticides. Insects that attacked cotton are able to survive despite the presence of gossypol in the glands on the plants. The cotton variety *Gossypium barbadense* was less affected than *G. hirsutum* when grown nearby in the Sudan as it has more glands with gossypol. A glandless cotton, without gossypol, selected to provide flour to make bread suitable for humans to eat succumbed to chafers (maize pests) in a trial in Zimbabwe. When a variety is selected for resistance to a pathogen, it may be successful for a limited period due to changes in the pathogen, so plant breeders need to select new hybrids continually to keep ahead of the pathogen.

Choice of crop variety continues to be an important component of IPM. Unfortunately market forces often dictate the growing of varieties, which are more palatable to humans and produce a higher, more profitable yield, rather than putting the emphasis on pest resistance. Nevertheless selection of crop varieties by traditional methods is essential alongside the new technology of genetic engineering to ensure that in emphasising one particular trait such as herbicide tolerance, the variety is still suitable for growing in different agro-ecological areas.

Present day pesticides

The agrochemical industry has changed significantly since the early days of pesticide development. While the older products no longer covered by patents have moved to generic companies, those investing in R and D have not only diversified into GM crops, but have realised that the registration authorities are unlikely to accept the most toxic pesticides, nor those which are very persistent in the environment. In Europe many of the older pesticides have been withdrawn as companies have declined to provide new data to meet the latest registration requirements. In some cases this has caused severe problems for some minor crops, for which no new pesticide has been registered, due to the small market for these crops. While broad-spectrum pesticides are needed with potential use of large areas of a major crop to cover the development costs, there has been recognition of the need for more selective products or more selective use of the broader spectrum chemicals.

Herbicides

Most new herbicides have been added to existing types, such as the sulphyl ureas. More refined production has led to the S-isomer of metolachlor replacing the earlier product. In general new herbicide products are often different combinations of herbicides in pre-mixed formulations to suit specific weed situations in different crops and countries. There has also been greater awareness of enhancing herbicide activity by recommending addition of certain adjuvants, such as methylated vegetable seed oils, that improve the spread of the spray deposit on foliage and increase the amount taken up by the weeds.

With the rapid increase in glyphosate-resistant weeds there is an urgent need to renew efforts to discover new herbicides utilising different modes of action and the development of other herbicide-tolerant traits in GM crops alongside integration with bioherbicides, cultural and mechanical methods for weed management (Green, 2014).

Fungicides

The strobilurins have been the main new group of fungicides, although new versions of older groups continue to be developed. Prothioconazole is a new azole (Mauler-Machnik et al., 2002). Ethaboxam is a new fungicide specific to controlling oomycetes such as grape downy mildew, and late blight on potato (Kim et al., 2002). Benzovindiflupyr is a new generation succinate dehydrogenase inhibitor (SDHI) which is now commercialized used to control several diseases and in conjunction with azoxystrobin to control soybean rust.

A detailed account of fungicides is given by Hewitt (1998).

Insecticides

Sparks (2013) reviewed the ongoing need for new insecticides due to the loss of existing products through the development of resistance, the desire for products with more favourable environmental and toxicological profiles, shifting pest spectra and changing agricultural practices. Groups of insecticides that included the pyrethroids and nicotinoids are now joined by the tetronic acids and diamides. Other new groups are chitin synthesis inhibitors and other insect growth regulators as well as the avermectins and certain other new pesticides.

Pyrethroids

The pyrethroids have been extensively used as a key replacement of the organochlorine insecticides, so reference to them is included here. In the 1930s and earlier the natural pyrethrins extracted from the flowers of *Chrysanthemum cinerarifolium* were widely used to control insect pests especially in dark warehouses, where sunlight could not break it down rapidly. As mentioned earlier, scientists wanted to develop a photostable version of the pyrethrins, and this was eventually accomplished with the development of permethrin and subsequently cypermethrin and deltamethrin (Elliott et al., 1973, 1978). Other pyrethroid insecticides have now been synthesised. They are reasonably persistent and although the active molecule can be highly toxic to mammals, the amount of pyrethroid actually applied is extremely small. Only a few grams per hectare compared with over 100 g or kg of older products. As with the earlier insecticides, once they became relatively inexpensive, they have been used too often, so selection of insects with resistance to them has occurred. It is, therefore, important in any one agro-ecosystem to limit the number of applications so only one generation of the pest is exposed to this group of insecticides. Even with use limited to a particular period as practised in Australia on cotton, the insects might still increase their tolerance to the insecticide if other tactics in an IPM programme are not followed. Thus the limitation of over-wintering populations is still important, for example where a closed season restricts the availability of host plants.

Neonicotinoids

A more recent development is this group of insecticides that emulate the effect of nicotine derived from tobacco. These insecticides, such as imidacloprid, act on the acetylcholine system by blocking the postsynapse nicotinergic acetylcholine receptors. The use of imidacloprid expanded rapidly both as a spray and seed treatment to utilise systemic activity against a range of pests including aphids and whiteflies. Similarly thiamethoxam is recommended for control of these pests. Another new neonicotinoid is clothianidin (Ohkawara et al., 2002). As discussed earlier, the EU decided on a moratorium on neonicotinoids used as seed treatments on flowering crops having assumed that their use was a cause of bee mortality. In reality other factors, including the presence of the Varroa mite and viral diseases as well a reduction in the availability of wild flowers in agroecosystems, were causing a decline in bee populations. Studies over a four-year period showed that colonies exposed to a crop sown with seed treated with thiamethoxam successfully overwintered and had a similar health status to control colonies in the following spring (Pilling et al., 2013). An important factor in using this group of insecticides has been the effectiveness of low doses applied as a seed treatment, thus avoiding sprays likely to drift into neighbouring areas with wild flowers. The debate on the impact of neonicotinoids is discussed by Godfray et al. (2014). If or when the ban on their use is lifted, there is the risk that their overuse will also lead to widespread resistance.

Phenylpyrazole

Fipronil is an example of another new type of broad spectrum insecticide used at very low dosages. It is also fairly persistent and has to be used with caution. It is nevertheless an important tool in some pest control situations as it provides a different mode of action.

Insect growth regulators (IGRs)

In contrast to the neurotoxic poisons, these chemicals interfere with the growth of the immature stages of insects. At the larval stage of an insect, it moults and forms a new skin or cuticle. This process is controlled by hormones and requires the production of chitin that forms the skin of the next larval stage. IGRs are in three main groups. These are the juvenile hormone analogues, anti-juvenile hormones and chitin synthesis inhibitors. The latter have been the most widely used and include diflubenzuron. Larvae affected by this insecticide will start to moult into the next instar but fail to complete the process. Action is slow as no effect is discernible until the insect moults. Adults are not killed, but there is some evidence that oviposition is adversely affected if they contact a sufficient dose. These insecticides have an extremely low toxicity to mammals.

Juvenile hormone analogues, such as methoprene have been used to control mosquito larvae and have also been successful in controlling Phaoroah's ants in buildings, although the effect is not seen for some time after the application. Tebufenozide is an example of the anti-juvenile hormone type, which cause larvae to form precocious adults.

Spinosads

In 1982, a new species of Actinomycete was found in a soil sample from the Caribbean. From this *Saccharopolyspora spinosa* two fermentation products led to the development of new class of insecticide, the naturalyte class. Spinosad is the first product to be commercialised in this class. Its mammalian toxicity and environmental profile make it an excellent insecticide in IPM programmes. Spinosad is degraded by sunlight but surface deposits become stabilized with activity at a range of pH values so it remains sufficiently effective on foliage to control a range of lepidopteran pests, yet it is safe to most beneficials. It has already been used extensively in conventional cotton crops to replace pyrethroids and to supplement control on Bt cotton.

Other insecticides

The quest for new compounds with different modes of action continues. An interesting new insecticide has been created by fusing the Australian funnel-web spider (*Hadronyche versuta*) venom with snowdrop flower (*Galanthus nivalis*) proteins (King and Hardy, 2012; Ikonomopoulou and King, 2013). The spider venom contains predominantly a disulphide-rich peptide neurotoxin which is active against a wide range of insect pests, although much large doses seem to be required to affect many natural enemies or bees (Windley et al., 2012). It is claimed that the novel insecticide, which has to be ingested affects the central nervous system of many insect pests, but bees were unaffected by doses higher than they would be exposed to in the field over seven days (Nakasu et al., 2014). An interesting development is to use the gene encoding ω-hexatoxin Hvla (Hvt) in cotton and tobacco plants. *Heliothis virescens* larvae mortality was 100% on transgenic tobacco, but not on cotton where the toxin expression level was much lower (Ullah et al., in 2015).

Pyridalyl is a new insecticide with low mammalian toxicity, yet good activity against lepidopteran pests (Saito et al., 2002). Another group, the spirocyclic phenyl-substituted tetronic acids is showing promise with spiromesifen having activity against whiteflies and spider mites (Nauen et al., 2002). Spirodiclofen has activity against mites and psyllids and scale insects offering an important tool in IPM fruit production (De Maeyer et al., 2002).

The diamides include an anthranilic diamide chlorantraniliprole and a phthalic diamide flubendiamide. (Satelle et al., 2008). Lahm et al. (2007)

reported that chlorantraniliprole has low mammalian toxicity but excellent activity against a range of lepidopteran species.

Biopesticides

While *Bacillus thuringiensis* (Bt is a bacterium, it is the toxin produced by certain genotypes that is very effective when ingested by certain groups of insects, as the alkaline conditions in the insect gut dissolves the protein crystal releasing the toxin. The spores are not toxic in humans as the pH of the gut is quite different. It is the gene that encodes for the toxin that has been incorporated into plant by genetic engineering, discussed earlier. *Bacillus thuringiensis* var *israelensis* (Bti) has an important role in controlling the larval stages of mosquitoes (vector of the malaria parasite) and black flies that are vectors of onchocerciasis (River blindness).

Apart from GM Bt crops, *Bacillus thuringiensis kurstaki* is also used as insecticides, but it is only really effective if deposited on the foliage being ingested by the younger larval stages. There is no contact activity. It is an important insecticide in forestry and organic agriculture, hence concern by some people about its wider use in genetically engineered crops. *Bacillus pumilius* (QST 2808) is now marketed as a fungicide as it is effective against downy and powdery mildews and rusts. It is an economical biological partner for disease control and also provides an excellent fit in resistance management programmes.

A new broad spectrum biopesticide has been introduced in the USA using *Burkholderia* spp. Strain A 396, which has been shown to be active against a wide array of chewing and sucking insects and mites, due to novel modes of action of a diverse set of compounds it produces. De Costa et al. (2008) had earlier shown that *Burkholderia spinosa* to be a promising biofungicide against post-harvest diseases of bananas.

These are living organisms that are applied to crops in much the same way as chemical pesticides. The main groups of organisms used are fungi, viruses and entompathogenic nematodes. Following the locust plague in the late 1980s efforts were made to develop a biological control method. *Metarhizium acridum* isolated from a locust in Niger and was developed as a mycoinecticide that was effective if sprayed in arid desert conditions in Africa. As logistics dictate an ultra-low volume method of application against locust hoppers or swarms, the initial study with *Metarhizium* was to assess whether the spores could be formulated in oil and remain viable. Fortunately the spores are lipophilic, so formulation in oil was possible and small-scale tests showed that locusts were killed by mycosis once infected with the spores, although it took a few days before death occurred. Infected locusts however would stop feeding and could also infect other locusts. Later studies showed that this mycoinsecticide was selective against grasshoppers and locusts, so that in contrast to an organophosphate insecticide, the natural enemies, including birds were unaffected, so control continued whereas where natural enemies had been decimated by a broad spectrum insecticide, immigrant locusts were able to survive. *Metarhizium acridum* has also been used in Australia against plague locusts to avoid insecticide residues on pastures for 'organic' meat production.

Research has also shown that baculoviruses of insects can be effective, although there are problems of stability in sunlight and the need to deposit the virus so that it is ingested. Nuclear polyhedrosis viruses (NPVs) and granulosis viruses have been used in insect control. Purification of the insect viruses has been considered essential as crude extracts obtained from dead insects can contain other viruses that resemble pox-like viruses, but purification tends to make them more susceptible to ultra violet radiation in sunlight. Thus formulations of baculoviruses need to contain a sunscreen. Most success has been against forestry pests as delay in mortality and some damage has been acceptable in contrast to horticultural crops. In Scotland the pine beauty moth larvae were effectively controlled by baculovirus sprays, while in Canada the sawfly *Neodiprion sertifer* has been controlled by an NPV. Some research is examining the possibility of using baculoviruses as means of delivering a toxin, such as the venom from arthropods to increase the effectiveness of the virus and increase the speed of action.

Recently there has been interest in developing the venom from spiders as biopesticides. As indicated above, the venom contains predominantly a disulphide-rich peptide neurotoxin which is active against a wide range of insect pests, although much large doses seem to be required to affect many natural enemies or bees. One interesting development has been to combine recombinant fusion proteins containing neuroactive peptides/proteins with a 'carrier' protein that confers oral toxicity. Nakasu et al. (2014) have shown that Hv1a/GNA (*Galanthus nivalis* agglutinin), containing an insect-specific spider venom calcium channel blocker (v-hexatoxin-Hv1a) linked to snowdrop lectin (GNA) as a 'carrier', is an effective oral biopesticide towards various insect pests. The combination of Hv1a/GNA has been shown to be a potentially specific biopesticide, showing no adverse effects on the honeybee, *Apis mellifera*, an economically important pollinator, while being toxic to agronomically important insect pests. A possible reason for the lack of toxicity towards honeybees is that it degrades within the bee, preventing accumulation of the fusion protein, even if exposure is repeated.

Larger living organisms used as control agents are the entomopathogenic nematodes (EPNs), such as *Steinernema feltiae*. As with Bt, it is a toxin that causes death, having been carried into the insect by the nematode. The infective juveniles (IJs) are applied, usually in large volumes of water to control soil pests, such as the vine weevil (*Orthorhinus klugi*). Care is needed when using conventional spray equipment to avoid damaging the nematodes by sheer forces in the pump or nozzle and high temperatures caused by recirculating the spray several times through the pump. Careful distribution is needed to avoid leaving sections untreated as the nematodes will not move through dry soil. Research has investigated the addition of certain polymers to spray to permit EPNs to foliar pests, by keeping sufficient moisture on the IJs for a longer period.

As the major agrochemical companies have now purchased biopesticide companies, the latter now have the benefit of a much larger marketing organism, enabling more biopesticides to be integrated with an IPM programme, rather than attempt to stand alonein 'organic' crop protection. Companies also benefit from the more rapid and lower cost of development and registration of the biopesticide, although certain data is required, such as establishing the actual identity/strain of the organism being marketed as a biopesticide.

More selective applications

The continued supply of broad-spectrum insecticides has led to a need to consider more selective treatments where possible. Elsewhere attention has been drawn to the need to avoid downwind drift affecting water-courses and natural habitats by careful selection of spray nozzles and use of no-spray barriers. Using monitoring systems can assist in limiting the number of treatments and timing those that are needed to have the greatest impact on a pest population. When a spray can be directed at the most susceptible stage, often the first or second larval instars, the dose that needs to be applied can often be lower than that given on the label.

Traditionally most pesticide sprays are directed downwards on field crops. However, by angling nozzles forwards or backwards, sometimes in both directions with twin nozzles, it has been possible to improve deposition on the nearly vertical stems of cereal crops and improve the effectiveness of fungicide treatments. Early crop establishment is important so that the crop can compete with weeds so seed treatment will continue to be an important technique for protecting young seedlings. Patch spraying was referred to earlier as it is particularly relevant to herbicide application where specific weeds can be mapped and treatments localised accordingly.

In vector control, the treatment of bed-nets is a selective treatment as the mosquitoes are only exposed to deposits when attracted by the person sleeping under the net. Unfortunately if mosquitoes are active when people are not protected by the nets, disease transmission can still occur. Other uses of lures to attract insect pests to treated surfaces are discussed later.

In locust control, if locust hopper bands can be detected, it is possible to apply barriers of insecticide such that as the hoppers cross the treated vegetation or eat it, they accumulate a toxic dose. Prior to 1980s, dieldrin was used, but with the banning of this organochlorine, it is now possible to use fipronil in non-crop areas. The aim is to treat a barrier of about 100 m wide separated by a distance of at least 600 m or more between barriers. The area untreated allows survival of non-target organisms that may be affected by the insecticide. With this technique, the overall dose is less than 1 g of the insecticide per protected hectare. The application of a chitin synthesis inhibitor IGR, such as diflubenzuron against hoppers is also recommended. IGRs are more selective, but are slower in action.

Pheromones

There are many chemicals, now usually referred to as semio-chemicals, which modify behaviour through communication between organisms. Among these are kairomones that signal between different species, for example by attracting a natural enemy such as a parasitoid to its prey. This may be the odour emitted by a pest-damaged plant or in some cases by a chemical, which is also a pheromone. The pheromones most used in pest control are sex attractants. Among the lepidoptera, the volatile odour released by a virgin moth, for example is a strong attractant that guides male moths, often very long distances, towards the virgin moth, so that mating can take place. Pheromones are highly species specific and effective at incredible small quantities. They may consist of several chemicals or isomers in specific ratios, with a similar mixture, but a different ratio being used by a closely related species. They break down rapidly, especially in sunlight, so do not leave any residues.

Several approaches have been tried to utilise pheromones in insect control. One approach is to use the pheromone in traps to monitor pest populations. Catches can show when an infestation is starting. Insect numbers in a trap do not directly relate to the size of a pest population as the proportion of insects trapped may decrease with a higher pest population. Results from monitoring with traps may indicate when a spray or some other control tactic is needed or merely warn the farmer to monitor his crop by examining the plants.

A major technique to control pests is to release so much pheromone in the environment of a crop that males have great difficulty locating a female and mating – the mating disruption technique. This technique has been used to control pink bollworm on cotton in Egypt where 'twist-ties' – thin tubes containing the pheromone were tied to about 1 in 100 plants. Pheromone was released over several weeks through the plastic wall of the tube. Another mass disruption technique was to formulate the pheromone in microcapsules and the spray them on a crop.

A more recent pheromone application technology is to use an electrostatically charged powder in small traps. A moth entering the trap gets coated by the powder, which is then carried by the male moth, when it flies from the trap. This Exosex Auto Confusion system allows a much lower dose of pheromone to be used, as only about 25 dispensers/traps are required per hectare. Inside buildings, the auto confusion system can now be used for stored products pests and the clothes moth. The Exosect system is approved in the UK for codling moth (*Cydia pomonella*) control in apple orchards, and elsewhere in Europe. The pheromone disruption technique is highly selective so must be used in conjunction with other control measures needed in a crop, but the number of insecticide sprays is reduced.

Mass disruption of mating may well integrate with the adoption of GM crop technology as the method may reduce mating of insects with resistance to the toxin in for example Bt transgenic crops and supplement the impact of mating with susceptible insects outside the crop area.

Another approach is to use the pheromone as an attractant to an insecticide deposit, the 'lure and kill' technique. One successful example is the use of grandlure, the pheromone of the boll weevil (*Anthonomus grandis*), which is attractive to both males and females. Placing the pheromone with an insecticide on 'weevil sticks' placed around cotton fields attracts weevils emerging after the winter resting stage (from diapause) (Brashear, 1997). Lure and kill has been used successfully in several government control programmes against bollweevil, such as USA (Daxl et al., 1995; Plato and Plato, 1997; Villavaso et al., 2003), Nicaragua, Paraguay, and other mostly Latin American cotton growing countries and offers possibilities of minimising insecticide use. Tsetse flies have been effectively controlled by placing a plastic vial of octanol on screens of material dipped in insecticide. The odour of octanol is a strong attractant to get the flies to sit on the treated surface. Where there are herds of cattle, the direct treatment of the animal's skin with a pyrethroid has also been effective by using the animal as the attractant.

Most of the alternatives to pesticides considered here have only been successful in a limited area against specific pests. While there is scope for research to extend their application to other pests and crops, the relatively low cost of pesticides makes it difficult to justify additional costs and management skills needed to fully implement newer techniques. Much depends on the availability of pesticides at an acceptable cost to farmers and society. The reduction in active ingredients within the EU has stimulated more consideration of alternative technologies, but for the foreseeable future pesticides will still remain a key weapon in pest management. The last section in this chapter deals with a specific new technology, the use of genetically modified crops. While the public has been reluctant to accept the new technology, at least in Europe, the area of GM crops continues to increase in many areas of the world.

GM crops

The introduction of transgenic crops in some countries has faced a regulatory hurdle. As pointed out by Bradford et al. (2005) the costs of meeting regulatory requirements and market restrictions guided by regulatory criteria are substantial impediments to the commercialization of transgenic crops. With experience from long-accepted plant breeding methods, they proposed that regulatory emphasis should be on phenotypic rather than genomic characteristics once a gene or trait has been shown to be safe.

In addition to the transgenic approach, such as 'Golden Rice' engineered to contain beta carotene, to combat vitamin A deficiency affecting young children, there are new techniques of marker assisted breeding. As an example Luo and Yin (2013) have developed a new improved variety of rice, which has a semi-dwarf phenotype with improved lodging resistance, a greater harvest index and it can survive after 2 weeks of complete submergence without significant loss of viability. This was achieved by marker-assisted selection to pyramid the semi-dwarf gene sd1, submergence tolerance gene Sub1A, blast resistance gene Pi9 and bacterial blight resistance genes Xa21 and Xa27 in the new variety. Thus significant improvements can be achieved by genetic engineering without the connotations of moving a bacterial or herbicide tolerant gene into a plant. The important factor is that new varieties with drought resistance, salt tolerance and better nutritional characteristics will be needed to feed the growing population. If pest and disease resistance are not included in the phenotype, pesticides will still be crucial to protect the crops from yield losses.

Where GM technology has been accepted, the continued use of a narrow range of pesticides, notably glyphosate, selection of resistant weeds has become a major problem. Similarly with extensive use of varieties with the Bt toxins, some insect pests are now tolerant. This has been evident particularly with the maize root worm (Gassmann et al., 2014) The threat of selecting resistant insect pests was recognised early on and suppliers of GM seeds advocated farmers should sow an area of non-GM crop alongside to allow some insect pests to survive and reduce the selection of resistant pests (Bates et al., 2005). Some seeds have stacked genes with toxins from different Bt genes, but there is still the possibility that if the same variety is grown over a period of years, the pests will eventually be more resistant to that combination of toxins. A major problem with 'industrialised' farming is the extensive areas of one variety/crop and less attention, if any, given to rotations. Rotation of crops and use of different herbicides is crucial to reduce selection of weeds resistant to glyphosate or similar herbicide-tolerant crops. Greater adoption of IPM should enable a better balance between the need to maintain and increase crop production using both GM and non-GM crops (Nicholas et al., 2011). As pointed out by Park et al. (2011), it would be unwise to ignore transgenic crops as one of the tools to help meet aspirations for increased sustainable global development.

Similarly, sowing of GM oil-seed rape (Canola) expressing resistance to certain herbicides, especially glyphosate, has increased to over 75% of the world-wide area. This has decreased herbicide use by 40% as the new herbicide-resistant varieties require only one or two applications of the single broad-spectrum herbicide, whereas more traditional unmodified crops need several applications of herbicides, often applied as mixtures. Application of herbicides to GM herbicide resistant crops is after crop establishment so they can be applied more accurately to specific weed-infested areas, thus providing savings for the farmer and net benefits to the environment.

A meta-analysis of 147 original studies has shown that adoption of GM technology has reduced chemical pesticide use by 37%, increased crop yields by 22% and increased farmer profits by 68%. Klümper and Qaim (2014) also point out that yield gains and reduction in pesticide use have been larger for insect resistant crops than for herbicide-tolerant crops, and that yield and profit gains are higher in developing countries than in developed countries. They hope that the robust evidence of the benefits of growing GM crops revealed by the meta-analysis may increase public trust in the technology.

Perceptions and hopes for the future

The general public continues to be wary about pesticides as they are acknowledged to be poisons and are used in such small amounts that they are perceived to be dangerous. A viewpoint exemplified by the reports of the Pesticide Action Network (Craig, 2004). At the same time the public accepts the everyday danger of venturing out on crowded roads where accidents occur far too frequently. Furthermore, the fuel used in our vehicles would not be accepted as a pesticide. It contains chemicals that cause cancer in laboratory animals, causes fire when ignited and produces toxic gases (PM10 and smaller) that pollute the atmosphere. Such pollution is not confined to cities, but affects everywhere with extensive roads. Many injuries and deaths occur and there are major costs in exploring, refining and transporting the fuel worldwide. Nevertheless transport is needed so man has accepted the risks associated with the use of petroleum products. The change to using lead-free fuels is one response to the concerns of our polluted air, especially in built-up areas. However, recently the exhaust from diesel cars has received closer attention in relation to air quality in densely populated areas.

As with the Green Revolution of the 1950s, when farmers were encouraged to speed up food supply production and overcome the problems of famine in many parts of the world, there is a continuing need today for multi-disciplinary research and extension to develop suitable packages of crop variety, whether GM or not, with the combination of weed, disease and insect pest management tactics that provide farmers with a profitable crop. Sustainable agriculture will require judicious use of pesticides, and this can only be accomplished if the users are properly trained and the public educated to accept healthy food, even if minute traces of a pesticide can be detected. Regulatory authorities in many countries have now excluded most if not all of the highly toxic or highly persistent pesticides, and with the agrochemical industry withdrawing support for many of the older products, the number of pesticides now available has declined significantly in Europe. Efforts by FAO and others to harmonise the data requirements should lead more countries to follow the same principles and limit the number of active ingredients that can be used.

The European Union's Sixth Environmental Action Plan aimed at reducing the impacts of pesticides on human health and the environment. By the introduction of IPM as a key policy, many European countries have now a National Pesticide Strategy. Some of these policies have aimed at reducing the quantity of pesticides applied or area treated. A 50% reduction by weight is easily achieved if a new pesticide is applied that requires a tenth or less of the dosage of an older product. However, the new more active molecule could have a similar impact on the environment, hence the need at the registration stage to assess accurately those physical and chemical properties and its toxicity, which can influence the spread and impact of a given product once it has been applied to a crop.

An alternative idea has been to assess the frequency of applications on a crop, but this can be influenced by seasonal differences affecting the incidence and severity of pests. Farmers with knowledge of local conditions may try to apply a lower dosage, but sometimes if their forecast is wrong, they may need to repeat a dose, thus increasing the number of applications, although the total dose may still be less than the manufacturers has recommended. Frequency of application can be a guide in certain areas where only a few crops and pesticides are used. Ideally farmers would prefer fewer spray applications, as soil compaction will be less if there are fewer passages with the sprayer across the field. Seed treatments enable a reduction in sprays as a low dose is applied effectively to young seedlings and persists over the period of crop establishment. In some places using a modular system, tracks for the sprayer are retained, thus cropping is confined to definite beds.

With a wide range of crops, different products and seasonal variations in pest incidence, there is no simple formula for assessing whether pesticide reduction policies are effective. Overall the need is to refine biological markers to assess the variation in bird and other wild animal populations, bearing in mind these too are affected by meteorological factors. However much has been achieved with improvements in decision making to meet targets set by the supermarkets and retailers under crop assurance schemes. Better-qualified operators of pesticide application equipment, routine inspection of sprayers and more education of growers on choice of pesticide will enable good yields to be obtained at minimal cost. Integration of non-pesticide tactics needs to be encouraged but in most cases will be a secondary role minimising pesticide use but not eliminating their use entirely on every crop. When used judiciously only when really needed, pesticides will remain an

Registration	- Avoid Class I (WHO) pesticides
Label	- Endeavour to use lowest effective dose
Monitoring	- Inspect crop regularly, treat only when needed
Operator safety	- Choose equipment with engineering controls to minimise exposure - Wear appropriate personal protective equipment (PPE) - Ensure supply of water for washing
Application technology	- Correct choice of nozzle - Reduce drift - Calibrate equipment - Use Buffer zone (Follow LERAP) - Use other specialised systems where appropriate e.g. downward directed air assistance tunnel sprayer droplegs rotary nozzles closed transfer systems where possible - Ensure sprayer washings do not enter drains

Fig. 8.5 (Adapted from a diagram by Franklin Hall).

important tool in our IPM/ICM armoury. Figures 8.5 and 8.6 illustrate the processes of policy and information services linked with better application that can lead to more efficient use of pesticides within the context of IPM.

The dichotomy with IPM between the developing countries' poor resource farmers and developed countries is very wide. The latter are learning with Government restrictions to endeavour to reduce chemical pesticide inputs, while there is a clear need to improve management of farms in many other areas of the world. International organisations have put much emphasis on Farmer Field Schools (FFS) to empower farmers with more skills to improve their decision making. The problem is getting the appropriate information to a vast number of small scale farmers. Agrochemical companies have stewardship programmes, but the sheer number of small scale farmers means that these efforts often reach only a small proportion of farmers. Media attention to agriculture, consolidation of efforts and sustained inputs into better qualified extension services are all needed to get better training to reach farms. No doubt with the phenomenal spread of mobile phones, the next generation of farmers will benefit most from more dissemination of information by apps on mobile phones.

In addition to pesticides on crops, insecticides continue to be important in vector control of diseases. There is a new malaria action plan to defeat malaria by 2030 which makes the case for eliminating the scourge of malaria over the next 15 years ensuring no resurgence of the disease, with

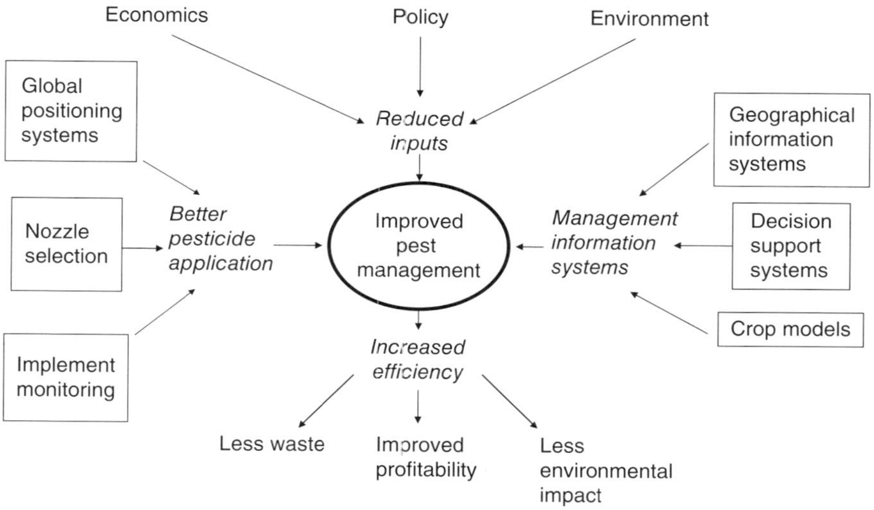

Fig. 8.6 Overview of improving integrated pest management (IPM) (Adapted from a diagram by Franklin Hall).

its associated crippling economic cost and avoidable deaths. Progress towards the new Sustainable Development Goals will be contingent on the continued reduction and elimination of malaria, but this will require heightened investment to expand efforts, stronger partnership – within and between sectors to develop new insecticides and integrated vector management programmes with a people-centred approach.

Eliminating malaria is critical to fighting poverty and improving maternal and child health, among others. In Africa alone, 10 000 women and between 75 000 and 200 000 infants are estimated to die annually as a result of malaria infection during pregnancy. It is unacceptable that the most vulnerable in our society are the least protected. Greater investment will be required in future generations, in the protection of mothers and their children.

References

Abedullah, K.S. and Qaim, M. (2015) Bt cotton, pesticide use and environmental efficiency in Pakistan. *Journal of Agricultural Economics* **66**, 66–86.

Bates, S.L. Zhao, J.Z., Roush, R.T and Shelton, A.M. (2005) Insect resistance management in GM crops: past, present and future. *Nature Biotechnology* **23**, 57–62.

Bradford, K.J., Van Deynze, A., Gutterson, N., Parrott, W. and Strauss, S.H. (2005) Regulating transgenic crops sensibly: lessons from plant breeding, biotechnology and genomics. *Nature Biotechnology* **23**, 439–444.

Brashear, A.L. (1997). *1996 SEBWEP Baitstick Utilization Summary*. USDA-APHIS Report. SEBWEF, Montgomery, AL.

Brooks, D.R., Bohan, D.A., Champion, G.T., Haughton, A.J., Hawes, C., Heard, M.S., Clark, S.J., Dewar, A.M., Firbank, L.G., Perry, J.N., Rothery, P., Scott, R.J., Woiwod,

I.P., Birchall, C., Skellen, M.P., Walker, J.H., Baker, P., Bell, D., Browne, E.L., Dewar, A.J.G., Fairfax, C.M., Barner, B.H., Haylock, L.A., Horne, S.L., Hulmes, S.E., Mason, N.S., Norton, L.R., Nuttall, P., Randle, Z.,Rossall, M.J., Sands, R.J.N., Singer, E.J. and Walker, M.J. (2003) Invertebrate responses to the management of genetically modified herbicide-tolerant and conventional spring crops. I. Soil-surface-active invertebrates. *Philosophical Transactions of the Royal Society London B* **368**, 1847–1862.

Champion, G.T., May, M.J., Bennett, S., Brooks, D.R., Clark, S.J., Daniels, R.E., Firbank, L.G., Haughton, A.J., Hawes, C., Head, M.S., Perry, J.N., Randle, Z., Rossall, M.J., Rothery, P., Skellern, M.P., Scott, R.J., Squire, G.R. and Thomas, M.R. (2003) Crop management and agronomic context of the FarmScale evaluations of genetically modified herbicide-tolerant crops. *Philosophical Transactions of the Royal Society London B* **368**, 1801–1818.

Craig, A. (2004) *People's Pesticide Exposures*. PAN, London.

Daxl, R., Ruiz Centeno, B. and Bustillo Caceres, J. (1995) Performance of the Boll Weevil Attract and Control tube (BWACT) in a 3 year area wide Nicaraguan Boll Weevil Control Program. Proceedings of the Beltwide Product Research Conference, National Cotton Council, Memphis, TN, USA.

De Costa, D.M., Zahra, A.R.F., Kalpage, M.D. and Rajapakse, E.M.G. (2008) Effectiveness and molecular characterization of *Burkholderia spinosa*, a prospective biocontrol agent for controlling postharvest diseases of banana. *Biological Control* **47**, 257–267.

De Maeyer, I., Peeters, D., Wijsmuller, J.M., Cantoni, A., Brueck, E. and Heibges, S. (2002) Spirodiclofen: a broad-spectrum acaricide with insecticidal properties: efficacy on Psylla pyri and scales Lepidosaphes ulmi and Quadraspidiotus perniciosus. The BCPC Conference Pests and Diseases, BCPC, Farnham, UK, pp. 65–72.

Elliott, M., Farnham, A.W., Janes, N.F., Needham, P.H., Pulman, D.A. and Stevenson, J.H. (1973) A photostable pyrethroid. *Nature* **246**, 169–170.

Elliott, M., Janes, N.F. and Potter, C. (1978) The future of pyrethroids in insect control. *Annual Reviews of Entomology* **23**, 443–469.

Gassmann, A.J., Petzoid-Maxwell, J.L., Clifton, E.H., Dunbar, M.W., Hoffmann, A.M. Ingber, D.A. and Keweshan, R.S. (2014) Field-evolved resistance by western corn rootworm to multiple *Bacillus thuringiensis* toxins in transgenic maize *Proceedings of the National Academy of Sciences* **111**, 5141–5146.

Godfray H.C.J., Blacquiere, T., Field, L.M., Hails, R.S., Petrokofsky, G. Potts, S.G., Raine, N.E., Vanbergen, A.J. and McLean, A.R. (2014) A restatement of the natural science base concerning neonicotinoid insecticides and insect pollinators. *Proceedings of the Royal Society B* **281**, 20140558.

Green, J.M. (2014) Current state of herbicides in herbicide-resistant crops. *Pest Management Science* **70**, 1351–1357.

Green, J.M. and Owen, M.D.K. (2011) Herbicide-resistant crops: utilities and limitations for herbicide-resistant weed management. *Journal of Agricultural and Food Chemistry* **59**, 5819–5829.

Heard, M.S., Hawes, C., Champion, G.T., Clark, S.J., Firbank, L.G., Haughton, A.J., Parish, A.M. Perry, J.N., Rothery, P., Roy, D.B., Scott, R.J., Skellern, M.P., Squire, G.R. and Hill, M.O. (2003) Weeds in fields with contrasting conventional and genetically modified herbicide-tolerant crops. I. Effects on abundance and diversity. *Philosophical Transactions of the Royal Society London B* **368**, 1819–1832.

Hewitt, H.G. (1998) *Fungicides in Crop Protection*. CAB International, Wallingford.

Hossain, F. Pray, C.E., Lu, Y., Huang, J. and Fan, C. (2004) Genetically modified cotton and farmers' health. *International Journal of Occupational and Environmental Health* **10**, 296–303.

Huang, J.K., Hu, R.F., Qiao, F.B., Yin, Y.H., Liu, H.J. and Huang, Z.R. (2015) Impact of insect-resistant GM rice on pesticide use and farmers' health in China. Science China Life Sciences, doi:10.1007/s11427-014-4768-1.

Ikonomopoulou, M. and King, G. (2013) Natural born insect killers: spider-venom peptides and their potential for managing arthropod pests. *Outlooks on Pest Management* **24**, 16–19.

James, C. (2012) *Global status of biotech/GM crops 2012*. ISAAA Briefs No. 44. ISAA, Ithaca, NY.

Jansen, A-E. (1999) Impacto ambiental del uso herbicidas en siembra directa. GTZ/MAG/DIA/DEAG, San Lorenzo. TFSC/SAR, Arusha. pp. 59–62.

Kim, D.S., Lee, Y.S., Chun, S.J., Choi, W.B., Lee S.W., Kim, G.T., Kang, K.G., Joe, G.H. and Cho, J.H. (2002) Ethaboxam: a new oomycetes fungicide. The BCPC Conference Pests and Diseases, BCPC, Farnham, UK, pp. 377–382.

King, G.F. and Hardy, M.C. (2012). Spider-venom peptides: structure, pharmacology, and potential for control of insect pests. *Annual Review of Entomology* **58**, 475–496.

Klümper, W. and Qaim, M. (2014) A meta-analysis of the impacts of genetically modified crops. *PLoS One* **9**(11), e111629.

Lahm, G.P, Stevenson, T.M., Selby, T.P., Freudenberger, J.H., Cordova, D., Flexner, L., Bellin, C.A., Dubas, C.M., Smith, B.K., Hughes, K.A., Hollingshaus, J.G., Clark C.E. and Benner, E.A. (2007) Rynaxypyr: A new insecticidal anthranilic diamide that acts as a potent and selective ryanodine receptor activator. *Bioorganic & Medicinal Chemistry Letters* **17**, 6274–6279.

Levie, A., Legrand, M-A., Dogot, P., Pels, C., Baret, P.V. and Hance, T. (2005) Mass releases of *Aphidius rhopalosiphi* (Hymenoptera: Aphidiinae), and strip management to control wheat aphids. *Agriculture, Ecosystems and Environment* **105**, 17–21.

Luo, Y. and Yin, Z. (2013) Marker-assisted breeding of Thai fragrance rice for semi-dwarf phenotype, submergence tolerance and disease resistance to rice blast and bacterial blight. *Molecular Breeding* **32**, 709–721.

Mauler-Machnik, A., Rosslenbroich, H-J., Dutzmann, S., Applegate, J. and Jautelat. (2002) JAU 6476 – a new dimension DMI fungicide. The BCPC Conference Pests and Diseases, BCPC, Farnham, UK, pp. 389–394.

Mensah, R.K. (1999) Habitat diversity: implications for the conservation and use of predatory insects of *Helicoverpa* spp. in cotton ecosystems in Australia. *International Journal of Pest Management* **45**, 91–100.

Nakasu, E.Y.T., Williamson, S.M., Edwards, M.G., Fitches, E.C., Gatehouse, J.A., Wright, G.A. and Gatehouse, A.M.R. (2014) Novel biopesticide based on a spider venom peptide shows no adverse effects on honeybees. *Proceedings of the Royal Society B* **281**, 20140619

Nauen, R., Bretschneider, T., Bruck, E., Elbert, A. Reckmann, U., Wachendorff, U. and Tiemann, R. (2002) BSN 2060: a novel compound for whitefly and spider mite control. The BCPC Conference Pests and Diseases, BCPC, Farnham, UK, pp. 39–44.

Neto, F.S. (1993) Controle de plantas daninhas atraves de coberturas verdes consorciadas com milho. *Pesquisa Agropecuaria Brasileira* **28**, 1165–1171.

Nicholas, A., Birch, E., Begg, G.S. and Squire, G.R. (2011) How agro-ecological research helps to address food security issues under new IPM and pesticide reduction policies for global production systems. *Journal of Experimental Biology* **62**, 3251–3261.

Ohkawara, Y., Akayama, A., Maysuda, K. and Andersch, W. (2002) Clothianidin: a novel broad-spectrum neonicotinoid insecticide. The BCPC Conference Pests and Diseases, BCPC, Farnham, UK, pp. 51–58.

Park, J.R., McFarlane, I., Phipps, R.H. and Ceddia, G. (2011) The role of transgenic crops in sustainable development. *Plant Biotechnology Journal* **9**, 2–21.

Pilling, E., Campbell, P., Coulson, M. and Tornier, I. (2013) A four-year field program investigating long-term effects of repeated exposure of honey bee colonies to flowering crops treated with thiamethoxam. *Plos One* **8** (10), e77193.

Plato, T.A. and Plato. J.C (1997) Alternative Uses of BWACT (Boll Weevil Attract and Control Tube) in Colombia, Brazil and Paraguay. *Proceedings of the Beltwide Cotton Production Research Conference*, National Cotton Council, Memphis, TN.

Ruiz, P., C. Novillo, J. Fernandez-Anero, M. Campos, (2001) Soil arthropods in glyphosate tolerant and isogenic maize lines under different soil/weed management practices. Proceedings of the 1st World Congress on Conservation Agriculture, Madrid, 1–5 October, 2001.

Saito, S., Isayama, S., Sakamoto, N., Umeda, K. and Kasamatsu, K. (2002) Pyridalyl: a novel insecticidal agent for controlling lepidopterous pests. The BCPC Conference Pests and Diseases, pp. 33–38.

Satelle, D.B., Cordova, D. and Cheek, T.R. (2008) Insect ryanodine receptors: molecular targets for novel pest control chemicals. *Invertebrate Neuroscience* **8**, 107–119.

Schuette, J. (1998) *Environmental Fate of Glyphosate: Environmental Monitoring and Pest Management*. Department of Pesticide Regulation, Sacramento, CA.

Sparks, T.C. (2013) Insecticide discovery: an evaluation and analysis. *Pesticide Biochemistry and Physiology* **107**, 8–17.

Squire, G.R., Brooks, D.R., Bohan, D.A., Champion, G.T., Daniels, R.E., Haughton, A.J., Hawes, C., Heard, M.S., Hill, M.O., May, M.J., Osborne, J.L., Perry, J.N., Roy, D.B., Woiwod, I.P. and Firbank, L.G. (2003) On the rationale and interpretation of farm scale evaluations of genetically modified herbicide-tolerant crops. *Philosophical Transactions of the Royal Society London B* **368**, 1779–1799.

Strandberg, B., Pedersen, M.B. and Elmegaard, N. (2005) Weed and arthropod populations in conventional and genetically modified herbicide tolerant fodder beet crops. *Agriculture, Ecosystems and Environment* **105**, 243–253.

Ullah, I., Hagenbucher, S., Alvarez-Alfageme, F., Ashfaq, M. and Romeis, J. (2015) Target and non-target effects of a spider venom, produced in transgenic cotton and tobacco plants. *Journal of Applied Entomology* **139**, 321–332.

Villavaso, E.J., Mulrooney, J.E. and McGovern, W.L. (2003) Boll weevil (Coleoptera: Curculionidae) bait sticks: toxicity and malathion content. *Journal of Economic Entomology* **96**, 311–321.

Windley, M.J., Herzig, V., Dziemborowicz, S.A., Hardy, M.C., King, G.F. and Nicholson, G.M. (2012) Spider-venom peptides as biopesticides. *Toxins*, **4**, 191–227.

Appendix 1 Standard terms and abbreviations

Technical terms

AAOEL acute acceptable operator exposure Level
Ach acetylcholine
Ache acetylcholinesterase
ADI acceptable daily intake
ADP adenosine diphosphate
AE acid equivalent
ai active ingredient
ALD_{50} approximate median lethal dose 50%
AOEL acceptable operator exposure level
ANOVA analysis of variance
AP alkaline phosphatase
approx approximate
ARC anticipated residue contribution
ARfD acute reference dose
as active substance
AST aspartate aminotransferase (SGOT)
ASV air saturation value
ATP adenosine triphosphate
BCF bioconcentration factor
bfa body fluid
BOD biological oxygen demand
bp boiling point
BSAF biota-sediment accumulation factor
BSP bromosulfophthalein
Bt *Bacillus thuringiensis*

Pesticides: Health, Safety and the Environment, Second Edition. G.A. Matthews.

Bti *Bacillus thuringiensis israelensis*
Btk *Bacillus thuringiensis kurstaki*
Btt *Bacillus thuringiensis tenebrionis*
BUN blood urea nitrogen
bw body weight
c centi-($\times 10^{-2}$)
°C degree Celsius (centigrade)
CA controlled atmosphere
CAD computer-aided design
CADDY computer aided dossier and data supply (an electronic dossier interchange and archiving format)
cd candela
CDA controlled drop(let) application
cDNA complementary DNA
CEC cation exchange capacity
cf confer, compare to
CFU colony forming units
ChE cholinesterase
CI confidence interval
CL confidence limits
cm centimetre
CNS central nervous system
COD chemical oxygen demand
CPK creatinine phosphatase
cv coefficient of variation
Cv ceiling value
CXL Codex Maximum Residue Limit (Codex MRL)
d day
DES diethylstilboestrol
DFR dislodgeable foliar residue
DMSO dimethylsulfoxide
DNA deoxyribonucleic acid
dna designated national authority
DO dissolved oxygen
DOC dissolved organic carbon
dpi days post inoculation
DRES dietary risk evaluation system
DT disappearance time
DT_{50} period required for 50% dissipation (define method of estimation)
DT_{90} period required for 90% dissipation (define method of estimation)
dw dry weight
DWQG drinking water quality guidelines
ε decadic molar extinction coefficient
EC_{50} effective concentration

ECD electron capture detector
ED_{50} median effective dose
EDI estimated daily intake
ELISA enzyme linked immunosorbent assay
e-mail electronic mail
EMDI estimated maximum daily intake
EPMA electron probe micro analysis
ERC environmentally relevant concentration
ERL extraneous residue limit
F field
F_o parental generation
F_1 filial generation, first
F_2 filial generation, second
FIA fluorescence immunoassay
FID flame ionization detector
FOB functional observation battery
fp freezing point
FPD flame photometric detector
FPLC fast protein liquid chromatography
g gram
G glasshouse
GAP good agricultural practice
GC gas chromatography
GC-EC gas chromatography with electron capture detector
GC-FID gas chromatography with flame ionization detector
GC–MS gas chromatography–mass spectrometry
GC-MSD gas chromatography with mass-selective detection
GEP good experimental practice
GFP good field practice
GGT gamma glutamyl transferase
GI gastro-intestinal
GIS geographical information system
GIT gastro-intestinal tract
GL guideline level
GLC gas liquid chromatography
GLP good laboratory practice
GM geometric mean
GMO genetically modified organism
GMM genetically modified micro-organism
GPC gel-permeation chromatography
GPPP good plant protection practice
GPS global positioning system
GSH glutathion
GV granulosevirus

h hour(s)
H Henry's Law constant (calculated as a unitless value) (see also K)
ha hectare
Hb haemoglobin
HCG human chorionic gonadotropin
Hct haematocrit
HDT highest dose tested
hL hectolitre
HEED high energy electron diffraction
HID helium ionization detector
HPAEC high performance anion exchange chromatography
HPLC high performance liquid chromatography
HPLC–MS high pressure liquid chromatography–mass spectrometry
HPPLC high pressure planar liquid chromatography
HPTLC high performance thin layer chromatography
HRGC high resolution gas chromatography
H_s Shannon–Weaver index
Ht haematocrit
I indoor
I_{50} inhibitory dose, 50%
IC_{50} median immobilisation concentration
ICM integrated crop management
ID ionization detector
IEDI international estimated daily intake
IGR insect growth regulator
im intramuscular
inh inhalation
ip intraperitoneal
IPM integrated pest management
IR infrared
ISBN international standard book number
ISSN international standard serial number
iv intravenous
IVF *in vitro* fertilization
IVM integrated vector management
k kilo
K Kelvin or Henry's Law constant (in atmospheres per cubic meter per mole) (see also H)[13]
K_{ads} adsorption constant
K_{des} apparent desorption coefficient
K_{oc} organic carbon adsorption coefficient
K_{om} organic matter adsorption coefficient
kg kilogram
L litre

LAN local area network
LASER light amplification by stimulated
LBC loosely bound capacity
LC liquid chromatography
LC–MS liquid chromatography–mass spectrometry
LC_{50} lethal concentration, median
LD_{50} lethal dose, median; dose letalis media
LCA life cycle analysis
LC_{Lo} lethal concentration low
LC–MS–MS liquid chromatography with tandem mass spectrometry
LD_{50} lethal dose, median; dosis letalis media
LD_{Lo} lethal dose low
LDH lactate dehydrogenase
LOAEC lowest observable adverse effect concentration
LOAEL lowest observable adverse effect level
LOD limit of determination
LOEC lowest observable effect concentration
LOEL lowest observable effect level
LOQ limit of quantification (determination)
LPLC low pressure liquid chromatography
LSC liquid scintillation counter
LSD least squared denominator multiple range test
LSS liquid scintillation spectrometry
LT lethal threshold
m metre
M molar
μm micrometer (micron)
MC moisture content
MCH mean corpuscular haemoglobin
MCHC mean corpuscular haemoglobin concentration
MCV mean corpuscular volume
MDL method detection limit
MFO mixed function oxidase
μg microgram
mg milligram
MHC moisture holding capacity
min minute(s)
mL millilitre
MLT median lethal time
MLD minimum lethal dose
mm millimetre
mol mol
MOS margin of safety
mp melting point

MRE maximum residue expected
mM milimoles
MRL maximum residue level
mRNA messenger ribonucleic acid
MS mass spectrometry
MSDS material safety data sheet
MTD maximum tolerated dose
n normal (defining isomeric configuration)
NAEL no-adverse-effect level
nd not detected
NEDI national estimated daily intake
NEL no-effect level
NERL no-effect residue level
ng nanogram
nm nonometer
NMR nuclear magnetic resonance
no number
NOAEC no-observed-adverse-effect concentration
NOAEL no-observed-adverse-effect level
NOEC no-observed-effect concentration
NOED no-observed-effect dose
NOEL no-observed-effect level
NOIS notice of intent to suspend
NPD nitrogen-phosphorus detector or detection
NPV nuclear polyhedrosis virus
NR not reported
NTE neurotoxic target esterase
OC organic carbon content
OCR optical character recognition
ODP ozone-depleting potential
ODS ozone-depleting substances
OM organic matter
op organophosphorous pesticide
Pa Pascal
PAD pulsed amperometric detection
2-PAM 2-pralidoxime
pc paper chromatography
PC personal computer
PCV haematocrit (packed corpuscular volume)
PEC predicted environmental concentration
PEC_A predicted environmental concentration in air
PEC_S predicted environmental concentration in soil
PEC_{SW} predicted environmental concentration in surface water
PEC_{GW} predicted environmental concentration in ground water

PED plasma emission detector
pH pH value
PHED pesticide handler's exposure data
PHI pre-harvest interval
PIC prior informed consent
pic phage inhibitory capacity
PIXE proton induced X-ray emission
pKa negative logarithm (to the base 10) of the dissociation constant)
PNEC predicted no-effect concentration
po by mouth
P_{ow} partition coefficient between n-octanol and water
POP persistent organic pollutants
ppb parts per billion
PPE personal protective equipment
ppm parts per million
ppp plant protection product
ppq parts per quadrillion (10^{-24})
ppt parts per trillion (10^{-12})
PSP phenolsulfophthalein
PrT prothrombin time
PRL practical residue limit
PT prothrombin time
PTDI provisional tolerable daily intake
PTT partial thromboplastin time
QSAR quantitative structure–activity relationship
r correlation coefficient
r^2 coefficient of determination
RBC red blood cell
REI restricted entry interval
Rf retardation factor
RfD reference dose
RH relative humidity
RL_{50} median residual lifetime
RNA ribonucleic acid
RP reversed phase
rpm rotations per minute
rRNA ribosomal ribonucleic acid
RRT relative retention time
RSD relative standard deviation
s second
SAC strong adsorption capacity
SAP serum alkaline phosphatase
SAR structure/activity relationship
SBLC shallow bed liquid chromatography

sc subcutaneous
sce sister chromatid exchange
SD standard deviation
se standard error
SEM standard error of the mean
SEP standard evaluation procedure
SF safety factor
SFC supercritical fluid chromatography
SFE supercritical fluid extraction
SIMS secondary ion mass spectroscopy
SOP standard operating procedures
sp species (only after a generic name)
SPE solid-phase extraction
SPF specific pathogen free
spp subspecies
sq square
SSD sulphur-specific detector
SSMS spark source mass spectrometry
STEL short-term exposure limit
STMR supervised trials median residue
t tonne (metric ton)
$t_{½}$ half-life (define method of estimation)
$\mathbf{T_3}$ tri-iodothyroxine
$\mathbf{T_4}$ thyroxine
TADI temporary acceptable daily intake
TBC tightly bound capacity
TCD thermal conductivity detector
$\mathbf{TC_{Lo}}$ toxic concentration, low
TID thermionic detector, alkali flame detector
$\mathbf{TD_{Lo}}$ toxic dose low
TDR time-domain reflectrometry
TER toxicity exposure ratio
$\mathbf{TER_1}$ toxicity exposure ration for initial exposure
$\mathbf{TER_{ST}}$ toxicity exposure ration following repeated exposure
$\mathbf{TER_{LT}}$ toxicity exposure ration following chronic exposure
tert tertiary (in a chemical name)
TEP typical end-use product
TGGE temperature gradient gel electrophoresis
TIFF tag image file format
TLC thin-layer chromatography
Tlm median tolerance limit
TLV threshold limit value
TMDI theoretical maximum daily intake
TMRC theoretical maximum residue contribution

TMRL temporary maximum residue limit

TOC total organic carbon

Tremcard Transport emergency card

tRNA transfer ribonucleic acid

TSH thyroid stimulating hormone (thyrotropin)

TWA time weighted average

UDS unscheduled DNA synthesis

UF uncertainty factor (safety factor)

ULV ultra low volume

UV ultraviolet

v/v volume ratio (volume per volume)

WBC white blood cell

wk week

wt weight

w/v weight per volume

w/w weight per weight

XRFA X-ray fluorescence analysis

yr year

$<$ less than

$\leq$ less than or equal to

$>$ greater than

$\geq$ greater than or equal to

Organisations and publications

ACPA American Crop Protection Association

ASTM American Society for Testing and Materials

BA Biological Abstracts (Philadelphia)

BART Beneficial Arthropod Registration Testing Group

CA Chemical Abstracts

CAB Centre for Agriculture and Biosciences International

CAC Codex Alimentarius Commission

CAS Chemical Abstracts Service

CCFAC Codex Committee on Food Additives and Contaminants

CCGP Codex Committee on General Principles

CCPR Codex Committee on Pesticide Residues

CCRVDF Codex Committee on Residues of Veterinary Drugs in Food

CE Council of Europe

CIPAC Collaborative International Pesticides Analytical Council Limited

COREPER Comite des Representants Permanents

EC European Commission

ECB European Chemical Bureau

ECCA European Crop Care Association

ECDIN Environmental Chemicals Data and Information Network of the European Communities
ECDIS European Environmental Chemicals Data and Information System
ECE Economic Commission for Europe
ECETOC European Chemical Industry Ecology and Toxicology Centre
ECLO Emergency Centre for Locust Operations
ECMWF European Centre for Medium Range Weather Forecasting
ECP Expert Committee on Pesticides
ECPA European Crop Protection Association
EDEXIM European Database on Export and Import of Dangerous Chemicals
EHC (number) Environmental Health Criteria (number)
EINECS European Inventory of Existing Commercial Chemical Substances
ELINCS European List of New Chemical Substances
EMIC Environmental Mutagens Information Centre
EPA Environmental Protection Agency
EPO European Patent Office
EPPO European and Mediterranean Plant Protection Organisation
ESCORT European Standard Characteristics of Beneficials Regulatory Testing
EU European Union
EUPHIDS European Pesticide Hazard Information and Decision Support System
EUROPOEM European Predictive Operator Exposure Model
FAO Food and Agriculture Organisation of the UN
FERA Food and Environment Research Agency
FOCUS Forum for the Co-ordination of Pesticide Fate Models and their Use
FRAC Fungicide Resistance Action Committee
GATT General Agreement on Tariffs and Trade
GAW Global Atmosphere Watch
GIFAP Groupement International des Associations Nationales de Fabricants de Produits Agrochimiques
GCOS Global Climate Observing System
GCPF Global Crop Protection Federation (formerly known as GIFAP)
GEDD Global Environmental Data Directory
GEMS Global Environmental Monitoring System
GIEWS Global Information and Early Warning System for Food and Agriculture
GRIN Germplasm Resources Information Network
HRAC Herbicide Resistance Action Committee
IARC International Agency for Research on Cancer
IATS International Academy of Toxicological Science
IBT Industrial Bio-Test Laboratories
ICBB International Commission of Bee Botany
ICBP International Council for Bird Preservation

ICES International Council for the Exploration of the Seas
ICPBR International Commission for Plant–Bee Relationships
ILO International Labour Organisation
IMO International Maritime Organisation
IOBC International Organisation for Biological Control of Noxious Animals and Plants
IPCS International Programme on Chemical Safety
IRAC Insecticide Resistance Action Committee
IRC International Rice Commission
ISCO International Soil Conservation Organisation
ISO International Organisation for Standardisation
IUPAC International Union of Pure and Applied Chemistry
JECFA FAO/WHO Joint Expert Committee on Food Additives
JFCMP Joint FAO/WHO Food and Animal Feed Contamination Monitoring Programme
JMP Joint Meeting on Pesticides (WHO/FAO)
JMPR Joint Meeting on the FAO Panel of Experts on Pesticide Residues in Food and the Environment and the WHO Expert Group on Pesticide Residues (Joint Meeting on Pesticide Residues)
NATO North Atlantic Treaty Organisation
NAFTA North American Free Trade Agreement
NCI National Cancer Institute (USA)
NCTR National Centre for Toxicological Research (USA)
NGO non-governmental organisation
NTP National Toxicology Programme (USA)
OECD Organisation for Economic Co-operation and Development
OLIS On-line Information Service of OECD
PAN Pesticide Action Network
RNN Re-registration Notification Network
RTECS Registry of Toxic Effects of Chemical Substances (USA)
SAICM Strategic Approach to International Chemicals Management
SCPH Standing Committee on Plant Health
SETAC Society of Environmental Toxicology and Chemistry
SI Systeme International d'Unites
SITC Standard International Trade Classification
TOXLINE Toxicology Information On-line
UN United Nations
UNEP United Nations Environment Programme
WCDP World Climate Data Programme
WCP World Climate Programme
WCRP World Climate Research Programme
WFP World Food Programme
WHO World Health Organisation
WTO World Trade Organisation
WWF World Wildlife Fund

Appendix 2
Checklist of important actions for pesticide users

Drift reduction

- Choose the nozzle carefully – coarse spray quality?
- (*Check suitability of nozzle for applying pesticide to the crop/weed.*)
- Consider using an air-induction nozzle.
- Change to a coarser spray for the last downwind swath.
- Use downwardly directed air assistance in arable crops.
- Check that boom height is not too high.
- Avoid too fast a forward speed.

Downwind protection

- Check wind speed: avoid too high a wind or very turbulent wind conditions, but also avoid very still conditions.
- Use the lowest effective dosage.
- Use buffer zones to protect sensitive areas.
- Use a hedge/windbreak: ensure that it has appropriate porosity and protect it with a buffer zone.
- Use another tall crop to act as a filter.

Water protection (in addition to the aforementioned text)

- Do not prepare spray on a hard surface.
- Collect any spillage.
- Only mix the right amount of pesticide for a spray treatment.

Pesticides: Health, Safety and the Environment, Second Edition. G.A. Matthews.

- Wash the sprayer tank in the treated field and use a diluted spray on the last swath.
- Wash the outside of the sprayer alongside a biobed.
- Use a biobed to dispose of any diluted pesticide.

Operator protection

- Always wear overalls. If not available, wear long trousers and long-sleeved shirt. (*Keep this set of clothes separate from normal washing.*)
- Choose the least toxic product which is effective against the pest/pathogen or weed.
- Wear gloves when preparing the spray and when there is potential exposure of the hands to pesticide.
- Wear a hat.
- Wear a face shield when preparing a spray.
- Wear an apron to protect overalls during preparation of a spray. (*Remove apron before entering the tractor cab.*)
- Wear a respirator if applying a fog in an enclosed space. (*Always check the validity of the respirator filter.*)
- Use ear protectors if engine/fan noise is above 75 dB.
- Have water available for washing hands.
- Have drinking water also available. (*Keep it well protected from the pesticide, and only drink after washing the hands and face.*)
- Use a closed transfer system, if available.
- Launder any clothing exposed to a pesticide separately from other clothing; if disposable overalls have been worn, dispose of them through appropriate waste disposal service.

Index

Pesticides: Health, Safety and the Environment, Second Edition. G.A. Matthews.